The Myths of Human Evolution

The MYTHS of HUMAN EVOLUTION

Niles Eldredge   Ian Tattersall

Columbia University Press
New York   1982

Library of Congress Cataloging in Publication Data

Eldredge, Niles.
The myths of human evolution.

Includes index.
1. Human evolution.   2. Social evolution.
3. Progress.   I. Tattersall, Ian.   II. Title.
GN281.E4   573   82-1118
ISBN 0-231-05144-1   AACR2

Columbia University Press
New York   Guildford, Surrey

Copyright © 1982 Columbia University Press
All rights reserved
Printed in the United States of America

Clothbound editions of Columbia University Press books are Smyth-sewn
and printed on permanent and durable acid-free paper.

Illustrations by Nicholas Amorosi

*The thing that hath been, it is that which shall be;
and that which is done is that which shall be done:
and there is no new thing under the sun.
Is there anything whereof it may be said,
"See, this is new"? It hath been already of old time,
which was before us.*

                        ECCLESIASTES 1:9–10

Contents

ONE   Of Myths and Mankind   *1*

TWO   Bestial Origins, Godlike Aspirations:
Mythic Views of Man's Place in Nature   *19*

THREE   Evolution: The Myth of Constant Adaptive Change   *29*

FOUR   Patterns Great and Small:
Evolutionary Change and the Fossil Record   *55*

FIVE   Fossils and Finders:
The Cast of Characters in Human Evolution   *67*

SIX   Patterns in Human Evolution   *119*

SEVEN   Patterns in History   *161*

EIGHT   Beyond Patterns: Theories of Change   *175*

Epilogue: Our Evolutionary Future?   *183*

Index   *189*

The Myths of Human Evolution

CHAPTER ONE

Of Myths and Mankind

THIS BOOK is about contemporary myths. The ancient Greeks, Norsemen, and Hebrews sang of heroic deeds, accounted for the origin of the universe as they knew it, and explained their own preeminence in it. These older myths, as pure stories, do not concern us. To the extent that they conflict with the findings of science and history on the origin and nature of things, these ancient myths have been laid to rest. Their value to us now resides in the insights they give us on the inner workings of the minds and cultures of ancient peoples—as well as the purely aesthetic values that good stories always retain.

The sorts of myths we have in mind are modern stories and attitudes about the origin and nature of our own biological species, *Homo sapiens*. Stories become myths when their truth has long been taken for granted. This is true in all walks of life, from superstitious injunctions against eating cookies hot out of the oven to some of our more precious scientific formulations. For science *is* storytelling, albeit of a special kind. Science is the invention of explanations about what things are, how they work, and how they came to be. There are rules, to be sure: for a statement to be scientific, we must be able to go to nature and assess how well it actually fits our observations of the universe. Science *is* theory, mental constructs about the natural world.

Some theories are better than others. Some have been tested more severely than others. When theories remain unexamined for a long time, they tend to take on mythic qualities. We are inclined to accept them as true, sometimes in the face of rather plain evidence to the contrary. Some of the myths we pursue in this book are of this sort: long-accepted scientific notions which do not stand up to close scrutiny.

Some of the most mythic of scientific notions lie in the realm of evolutionary biology. Evolution—the proposition that all organisms are related—is as highly verified a thesis as can be found in science. Subjected to close scrutiny from all angles for over a century now, evolution emerges as the only naturalistic explanation we have of the twin patterns of similarity and diversity that pervade all life. The basic notion that life has evolved is as certain as the existence of gravity or the idea that the earth is spheroidal. We call such highly verified notions "facts" when they consistently escape all attempts to prove them false. Evolution is no myth.

But *how* life has evolved is another matter entirely. Our standard expectation of evolution—slow, steady, gradual improvement, hence change, through time—is indeed a myth. Here is a scientific myth born of another myth more generally held by society at large in Victorian times: the myth of progress. The expectation that progress is inevitable paved the way for acceptance of a biological concept of evolution (which is good) and a specific notion of how that process works (which, it turns out, is not so good).

We are aware that theories posed as alternatives to their entrenched predecessors are themselves—if they are good enough—the stuff of future myths. This is not our concern; we can only deal with prevailing myth. As the anthropologist Marvin Harris in his book *Cannibals and Kings* has noted, the myth of progress stems from the myopia accompanying an unprecedented rise in general living standards once the Industrial Revolution finally harnessed the tremendous energy potential of fossil fuels. More typical cultural world views, like that of the Preacher of Ecclesiastes, stress the sameness of things. If change is predicted, it is the gloomy expectation of worsening circumstances. Progress still typifies our own (Western world) outlook, though it has taken some hard knocks recently. Even under its giddiest sway, the notion that things ain't what they used to be is always waiting in the wings, providing thematic counterpoint to the myth of progress in all things.

Biological evolution is unlikely to be accepted in a social circumstance where stability reigns as the world view. General notions of progress foster more particular notions of change and development: of the physical universe, among animals and plants, and within specific components of culture. But we shall see that the notion of progress, at least in biology, has been taken too far. Expectation colored perception to such an extent that the most obvious single fact about biological evolution—*non*change—has seldom, if ever, been incorporated into anyone's scientific notions of how life actually evolves. If ever there was a myth, it is that evolution is a process of constant change.

The data, or basic observations, of evolutionary biology are full of the message of stability. Change is difficult and rare, rather than inevitable and continual. Once evolved, species with their own peculiar adaptations, behaviors, and genetic systems are remarkably conservative, often remaining unchanged for several millions of years.

In this light, it is wrong to see evolution, or for that matter human history, as a constant progression, slow or otherwise. The history of the world is rather one of fits and starts, of new breakthroughs followed by rapid development, followed in turn by quiescence. New possibilities seem to have something of the quality of a vacuum: nature abhors an unexploited innovation. As soon as a new avenue is opened up, all of the possibilities inherent in it are quickly explored, and a new equilibrium is reached, a new status quo established. Subsequently, inertia will be on the side of that status quo, until something comes along with enough force to shake it up. We will look at various areas of human experience and see how what we know of them fits our view. That history repeats itself we all know; but that it repeats itself not only because human memory is short but also because there is simply a finite number of possibilities in the human experience we are less likely to realize as long as we are wed to the idea of inexorable progression, of inevitable change.

The myth of progression, that change of a positive sort is inevitable and constant, that the history of mankind is one of a struggle from brutishness to the niceties of modern societies, from barbarianism to sophistication, infects the general perception of cultural evolution. But the archaeological record of prehistory, as well as the histories of many of the early states, also bespeaks great stability, long-term persistence of cultural traditions. Stylistic traditions in art and artifacts—from hand axes to sculpture—can, and often do, persist for millennia. Again, the myth of constant, even rapid, cultural evolution is fostered mostly by the whirlwind of technological change in which we are currently (and perhaps only temporarily) enveloped. But historically, patterns of cultural change tend to show stability interrupted by occasional, usually rather rapid, change, rather than linear, constant modification for the better.

This might seem all too distressingly familiar to readers oppressed with the recent rash of books seeking to explain the cultural history of *Homo sapiens* in biological terms. So we note right here that the similarity in patterns of change in physical evolution and some aspects of cultural evolution is only that: a similar pattern of episodic change. Each was denied by those who over the past century have steadfastly maintained the myth of progressive change. We shall have far more to say, in the next chapter, about specific proposals (such as those of sociobiology), but here we simply state the obvious: similarity in pattern need not imply common causes underlying the patterns. Inheritance is a crucial concept in any idea of change. Anatomies are inherited genetically. Some behaviors, such as nest building in birds, are largely, or even wholly, inherited genetically. Attributes of human cultures are transmitted by learning, though some psychological attributes of *Homo sapiens* seem ubiquitous within the species, suggesting the possibility of a degree of genetic control of some basic aspects of behavior. The human capacity to learn allows an "inheritance of acquired characters" in cultural evolution not encountered (to any significant degree) in biological evolution. Cultural evolution, theoretically, proceeds in a different fashion and, potentially at least, quite a bit faster than physical evolution. The two are by no means the same—which makes the superficial similarity of the patterns they each produce an interesting intellectual challenge that we will approach in chapter 8. But in no sense can we explain human cultural evolution in terms of biological evolution. We will deal with this issue in far greater detail in the next chapter. But first we will outline our basic views on human physical and cultural evolution, then lay yet another modern myth to rest.

Human Evolution—The Beginnings

The band of ape-men had been at the water hole for some time. While a few of the more robust males stood guard, the rest of the group of twenty or so drank. Suddenly, another band of ape-men appeared. They looked just like those already at the water hole. Jumping up and down, darting forward and then retreating, the newcomers bullied and threatened. With growls and screams, the first band stepped back from the water. The competitors had prevailed, and they now could settle in to drink their fill.

The scene shifts, and some days later we see a young ape-man alone amid a pile of bones in a rocky wasteland. Toying aimlessly with the bones, he picks up the long shaft of a giraffe leg. Waving it about, he accidentally strikes a rib cage lying nearby. The rib cage explodes, shattering into a thousand pieces. The ape-man stares in wonder at the leg bone still clutched in his hand. What has he done? He tries it again, this time smashing a string of neck bones into smithereens. With sudden glee he strikes out again and again in all directions, delighted as the significance of his discovery dawns on him.

Some days later we see again a band of ape-men at the water hole. Creeping up upon them is another, nearly identical band. But this time there is a difference; some of the members of the band stealthily nearing the water hole are carrying bone weapons. Revenge will be sweet, and sudden and merciless as well. The two bands confront each other, and again yells and screams fill the air. Then one of the attackers darts forward, waves his club, and dashes back. Another comes out brandishing a bone. A bit bolder, he runs up closer. The other ape-men stand their ground at the water hole, relying on their yells and screams to drive off the intruders. After all, it had worked before; that was how they had won possession of the water hole in the first place. But this time it fails to work. With a mighty yell, one ape-man brings his club crashing down on the skull of an unsuspecting rival. The bewildered band retreats from the water, stunned by the sudden demise of one of its leaders. The second band, the original possessors of the water hole, had triumphed, thanks to the invention of crude but effective weaponry.

This vignette from truly ancient history may or may not have happened. But based on the effective screenplay of Stanley Kubrick and Arthur C. Clarke, in the early scenes of the movie *2001*, it makes an absorbing scenario. It may also present us with an accurate view of the basic patterns of change in human evolution.

Kubrick and Clarke based their imagery on the competition between rival groups. According to their view, change should be stepwise and sporadic. First, an individual invents something. The rest of the group learns it. The advantage is then used to the benefit of the entire group, in this case, by routing the unfortunates of the competing band who did not share in the invention. Here we see at work a mechanism which gives us a picture not of gradual, progressive improvement, but one of sudden invention, immediate impact, and, presumably, a long wait until something new is invented. Then the nature or the quality of life will suddenly be changed again.

The Kubrick–Clarke scenario deals with behavior and implements, two ingredients of cultural change; and, as we will see shortly, it fits very well the pattern we find in the archaeological record. But Kubrick and Clarke might just as well have dwelt on mankind's physical evolution. For the same possibilities of different styles of change appear here as well. Is anatomical change in human—or any other—evolution sudden and rapid, followed by long periods of little or no change? Or do we find a pattern of gradual, progressive evolution among our precursors, from one state to another—from ape-man to modern *Homo sapiens*? Popular belief and anthropology texts alike agree that slow, even change over millions of years gradually transformed the ape-man of the African savanna into our present selves; but let us look briefly at the fossil record itself.

The details of the earliest evolution of the human line are still very vague, and as we shall see later, the time at which the human line branched off from that leading to our closest relatives, the African apes, is still shrouded in obscurity. But it is now well known that by 3 million to 4 million years ago, in Africa, there lived upright bipeds who, if not yet qualifying to be known as human, nonetheless possessed some of those attributes which are unique to our own species in the modern world, and who can at the very least be admitted to our own zoological family, Hominidae. Perhaps the most dramatic evidence of the bipedality of this early form is the set of footprints that have been found at Laetoli, in Tanzania. Well over 3 million years old, the tracks are sharply imprinted into volcanic ash, and show the direct, purposeful stride of a small, but fully erect, early man. All the characteristics of our own footprints on a wet beach are there, frozen in stone. The great toe is in line with the others, not divergent from them; the foot was sprung on arches. Mankind was up and walking close to 4 million years ago, and quite possibly a good bit earlier than that.

The early African hominids who produced these tracks were small, lightly built creatures. Their teeth were large relative to the size of their bodies, but their brains were small. They lived in an open environment, vulnerable to predators such as the large cats; but their way of life must have been a successful one, for they remained largely unchanged for at least 2 million years. In this time they were joined by other hominid species: one, for instance, of larger size and more robust build, and another still lightly built but with a rather larger brain. The remains of each of these new species appear in the fossil record rather abruptly. And later, at about 1.6 million years ago in Africa, and conceivably in Asia even earlier, we find, again unanticipated, yet another kind of early man, larger brained still, closer yet to ourselves. And in each case where a good enough fossil record exists, we find these species enduring through long periods of time, the history of each of them marked less by change than by stability.

So the pattern emerges. We do not see constant progressive brain enlargement through time, or a climb to a more completely human posture. We see instead new "ideas," like upright posture, developed fully from the outset. We see the persistence, through millions of years, of species which continue on unchanged for as long as their environment remained essentially unchanged. We see change coming mostly through the origin of entirely new species: new, independent reproductive groups. Their sudden appearance alongside their unchanged ancestors reflects a common pattern in the geography of evolution. Throughout the animal world new species typically evolve, usually rather rapidly, near the remote edges of the domain of the ancestral species. If the new species survives, it may one day take its place in the sun alongside its ancestor, or even completely eclipse it. Such seems to have happened with these early ape-men and their descendants. And this picture of stability for long periods, interrupted by abrupt change, is typical of the fossil record of all life. Human evolution is no different in style from the evolution of any other group of related animal species.

Man the Toolmaker

People have realized at least since classical times that man's propensity for making tools sets him apart from the rest of nature. And even though we know today that toolmaking and use, in the strictest senses, are not our exclusive province, it is nonetheless clear that the complexity of our technology, even as it is expressed in the earliest of human societies, is altogether unique. Certainly, chimpanzees strip and prepare twigs to "fish" for termites in their earthen mounds. Capuchin monkeys use stones to crack open hard nuts. Baboons kill scorpions with rocks before removing their stings and eating them; and even otters use stones to beat open shellfish. But this gray area, much as has been made of it, is largely of academic interest. Man alone is distinguished not only by the richness and variety of the tools he makes—and of the things he makes with those tools—but by the fact that he has become dependent on his tools for his very survival.

Making tools and using them to manipulate the environment: these human propensities are intimately bound up with our way of life and in numerous and subtle ways even with our image of ourselves. And living as we do in a period of history when every day seems to bring us news of yet another development or refinement of our already immensely elaborate technology, it seems natural for us to think of the development of this technology upon which we depend as a gradual, almost inevitable progression from its simplest beginnings. But while this might appear to be intuitively reasonable, it would also be mistaken. For just as we saw has been the case with our physical evolution, we find that the process of technological development has been one of fits and starts. As we look back into prehistory, into the unfolding record of human cultural achievement, we find the same pattern repeating itself over and again. The archaeological record shows us that once a new development—a breakthrough—appeared, there was invariably a rapid exploration of all the possibilities it opened up, followed by a fallow period of waiting for the next breakthrough to occur. Like nature itself, mankind has from the earliest times been opportunistic, quick to pick up on new openings once they were there. But in the past no less than today, truly creative leaps were much rarer than the exercise of human ingenuity in exploiting them.

Stone tools constitute the earliest evidence we have of the exercise of the human capacity for material culture. And these appear quite suddenly in the archaeological record, at around two and a half million years ago. Of course, it seems reasonable enough to assume that the earliest attempts of our remote progenitors to use or to fashion implements for digging, to aid in the gathering of wild vegetable foods, or perhaps according to the Kubrick-Clarke scenario, were made in materials a little kinder than unyielding stone. Wood, at least in the first case, is the obvious candidate. But wood and soft materials of that sort do not preserve except under the most unusual conditions, so we can only speculate about the kind of material culture that may have preceded the invention of the indestructible stone tool. However, what is most fascinating—and instructive—about the earliest stone tools is that from their first appearance they existed in the form of a tool *kit*. The various components of this tool kit were crudely made, to be sure, with flakes struck off a stone core to produce a cutting surface; but a variety of such shapes is recognizable, and along with these different forms, different functions—choppers, scrapers, and awls, for instance. So once the concept of the stone tool was invented, once our precursors learned to envision the form of a tool within a raw piece of stone and to bash that stone with another until that form was realized, it seems that all the basic variants of form and function made possible by that crude technique were invented virtually instantaneously.

The fact that tool kits of this kind continued to be made for well over another million years without much refinement shows that all the avenues opened up by this revolutionary concept were fully explored from the outset. This early stoneworking tradition is known as the Oldowan, from the famous Olduvai Gorge of Tanzania, where it has been most intensively studied. It made its first appearance at the gorge around 1.8 million years ago, and survived there in the form of the "Developed Oldowan" until some 800,000 years ago. Several other tool-making traditions from elsewhere in Africa seem to be no more than variants on the same theme; they differ more in the proportions of the different tool types represented in the kit than in the kinds of tool produced. And essentially, there was no change in the method of producing them.

But suddenly at Olduvai, after about half a million years of unchanging technique in the production of stone tools, a completely new tradition of stoneworking appeared. This industry, known as the Acheulean, depended upon a revolutionary new concept in toolmaking. In the Oldowan tradition, as we have seen, a stone core was bashed until it conformed to a certain shape. In the new technology a core was prepared too. But then a long flake would be struck off it; this was the tool, and the core itself was discarded. This change in concept permitted the addition of a whole panoply of new tools to the kit, and from the first exercise of this technique we can recognize hand axes, cleavers, knives, picks, and so forth, as well as tools already familiar for hundreds of millennia.

And once again, this spurt in technological achievement was succeeded by a period of relative equilibrium. It is, after all, the fate of all successful revolutions to become the status quo. Refinement, of course, did occur; and man of the Middle Stone Age, as this period of our prehistory is called, rapidly became capable of producing, with the minimum of effort, a comparatively limited range of tools to astonishingly uniform specifications, from an enormous variety of different stony materials. But, as before, it was hundreds of thousands of years before another advance was made that was comparable to the one which had heralded the beginning of the Middle Stone Age.

This pattern was repeated again, as we will see later. It is enough to establish here that essentially the same pattern of change has held good in both the physical and the cultural evolution of mankind. This may seem remarkable on the face of it, since the mechanisms of change are different in each case. But this difference is reflected elsewhere, in the fact that, while physical and cultural change have both occurred in the same stepwise manner, the two have been out of phase; there has been no necessary relationship between the introduction of a new kind of technology and the evolution of a new species of proto-man. In any event, what we find in both areas fails to conform to the presumption of slow, steady change in which we have been taught to believe. It is vital for us to grasp the essentially episodic nature of change, not only so that we may better understand how we got to where we find ourselves today, but in order that we may have a rational basis for understanding where our species is going.

How We Study Ourselves: Reductionism and Other Myths

The American public has an ambivalent attitude toward science. The overly romanticized picture of the intense scientist scrutinizing test tubes as retorts bubble in the background quickly turns to images of madmen obsessed with arcane experiments which will give them control over society. Science seems, somehow, a necessary evil. Without it, we would not have our telephones and TV sets—nor would we have the neutron bomb.

The "positive" side of such ambivalence tends to see scientists as bright but dispassionate automata who care for little but their esoteric research. Cold, calculating objective reality is the goal, and physicists immediately spring to mind when one thinks of scientists. While we remember the unkempt Albert Einstein clad in a baggy sweater, the white-draped figure in the sterile laboratory performing his carefully controlled experiments is the quintessence of our general view of the "scientist."

There is, of course, some substance to this view of science and scientists. Physicists deal with force, energy, matter. It is they who investigate, both theoretically and experimentally, the very building blocks of matter. In search of *the* elemental particles, physicists anatomize atoms, getting closer and closer to making truly general statements about the nature of all matter. For, while all atoms are alike in organizational plan, they differ significantly in detail. The search for more fundamental particles—below electrons, protons, and neutrons—is a search for properties of matter more general than atoms themselves.

Hence the two themes which combine to give us "physics envy." First we see the physicist as the paragon of what a scientist ought to be; the height of objectivity with carefully controlled measurements, he is approached in his perfection by chemists, then by "applied" chemists—those involved in biochemistry and geochemistry. Other aspects of biology and geology, not to mention psychology and anthropology, fare less well. The phenomena addressed by these fields are larger and somehow messier. And the techniques employed to study, say, the evolution of mammals are far less meticulously precise, organized, or objective than the bubble chambers of a physicist's cyclotron. We get farther away from "hard" science the farther we go from physics.

The other component of physics envy is more subtle, but also more interesting. The notion that physicists actually study phenomena of more general importance than any other branch of science symbolizes a frame of mind we call "reductionism." A physicist is hardly to be faulted for seeking the ultimate particles comprising matter. But from such searches for progressively more general explanations the moral is often drawn that all science must work in this fashion. There is, in other words, a belief that discussion in more general terms is more "scientific" than are more specific analyses. In this way topics are said to be "reduced" to more general terms.

A standard example of reduction centers around the chemical compound "water," H_2O. Knowing the properties of the atoms hydrogen (H) and oxygen (O), both of which are gases at room temperature under normal pressures, in principle we should know what water would be like. Antireductionists point to the "emergent properties" of water—so unlike the gases which its two constituent elements form when uncombined. But proreductionists reply that, present ignorance aside, in principle the properties of H_2O would be predictable given sufficient knowledge of H and O. Despite the appeal of "emergent properties," reductionism pervades the minds of most active scientists today. Sometimes, it seems, we all wish we were physicists and that all of our explanations of the world were couched in physicists' language. Though it seems ludicrous to "reduce" human evolution to the terms of physics, most of us think that only pragmatic reasons prevent us from achieving this goal. We simply haven't a clue how to understand the distribution of differently shaped human bones in time and space in terms of atoms, let alone subatomic particles. But a reductionist interested in human evolution can do the next best thing: he can try, for example, to reduce human cultural evolution to the principles of general biological evolution. He can reduce the principles of biologic evolution to the principles of genetics. By reducing genetics to the architectural and chemical properties of a few macromolecules (DNA and the various forms of RNA), he gets still closer. These gigantic molecules are made up of smaller chemical constituents (nucleic acids, for example). And all are made of atoms. Theoretically, *in principle*, a reductionist could explain human evolution in terms of the laws of physics. But failing that now, he can reduce one level of complexity to another, more general system: cultural evolution to general biological evolution, large-scale evolutionary phenomena to general genetics, and so forth.

But all this is to misread the lessons of the physicist in search of progressively more elementary particles. Physicists theorizing about these particles of matter hope to make statements true of *all* matter, be it arranged as free atoms or atoms combined into more complex systems. The various systems in which these particles are found are irrelevant to physicists. And here is the key objection to reductionism: differences between two systems are not explained by the properties common to both. Electrons and protons are in all atoms but also exist free of atoms. Electrons and protons (and neutrons, which occur in all atoms except hydrogen) are more general properties than are atoms. It is not the general properties of electrons, protons, and neutrons, but rather their number and arrangement that account for the differences between hydrogen, helium, and all the other elemental and isotopic forms of atoms.

Thus, to understand natural phenomena, we must seek to understand first how widely each phenomenon is distributed in nature. A statement about the nature of electrons that pertains solely to free electrons and not those zipping around the periphery of atomic nuclei is not a truly general statement about electrons. And it is simply no good to make a general statement about "life" if we include animals and plants but exclude fungi and the vast hordes of unicellular organisms.

Once we are sure the system is complete (is a "natural" one), the second step is to explain the differences between subunits within the system. The differences between plants, animals, and fungi do not flow from their common possession of complex ("eukaryotic") cells—but rather from the different metabolic mechanisms for obtaining and utilizing energy within these cells that each group has acquired as an evolutionary specialization. Joint possession of eukaryotic cells, in other words, unites plants, animals, and fungi (and some unicellular microorganisms as well) into one large, natural evolutionary group but tells us nothing about how and why plants, animals, and fungi differ.

Similarly, characterization of elements such as kinship patterns and religion, found in virtually all forms of human social organization, helps define the uniqueness of human society in the evolutionary realm but explains nothing about why religions, kinship patterns, and so forth differ in detail from group to group. Reducing the description of a system to the common elements that pervade it automatically prevents us from understanding the nature of the differences between elements within the system. To study such differences, we must factor out the common elements—*not* seek to explain the whole system solely in terms of its common elements. It boils down to this: you cannot explain the differences between two things by talking only about the similarities between them. Viewed this way, reductionism—harping only on the shared, common elements of a system—is useless as a strategy to explain the history of change and differentiation within a system, be the problem the evolution of all of life or the history of human societies.

This view of nature is fundamentally hierarchical. It says that all natural phenomena have a finite distribution, some very broad (particles of matter), some very restricted (human social organization, limited to one species, *Homo sapiens*). All scientists, reductionists and antireductionists alike, seek to understand the distribution of properties in nature. All realize that some properties are more widely dispersed than others. The reductionists claim that the more generally distributed properties, when properly comprehended, explain the systems composed of less widely distributed properties. But the antireductionists, seeking to understand the differences between systems, feel that properties distributed in more than one system are irrelevant to understanding the differences between systems.

Our position, as myth debunkers, is that reductionism has been followed with knee-jerk regularity in most areas of science to the present day. Physics envy, as one manifestation of reductionism, is not particularly funny. Given its general acceptance, reductionism borders on myth.

But we caution against the opposite tendency to identify a system, factor out all components found elsewhere, and claim automatically that important processes not found elsewhere are acting in the system. All this will become clearer when we take a look at the "man-as-an-animal" vs. "man-as-culture-bearer" theme in the next chapter. This is often seen as an either/or argument, reductionists explaining all human behavior in terms of biological principles (themselves reduced to genetics) and some opponents insisting that emergent properties beyond the biological realm exclusively govern human behavior. The real problem, however, is to take each phenomenon and see how widely it is distributed in nature. As we shall see, there is more to the antireductionist claim that human cultural behavior is unique to our own evolutionary lineage than reductionists, like sociobiologists, would have us believe.

Thus we have several kinds of myths in mind. Some are substantive: specific notions about how life in general evolves, how our own species has evolved, and the basic nature, and patterns of change, of culture. We will compare each myth with what is known today that bears on the issue, and we will then present another explanation about what happened and why. In so doing, we will not be idly speculating, for the rules of the game are clear: scientific statements must be put in such a way that their veracity can be called into question. The explanations we give in this book, while appearing to us to be superior to available myths, could also be proven less satisfactory than some ideas as yet unconsidered. To be scientific, ideas must be criticizable. To be the stuff of new myths, they have to be damn good explanations, and the jury has only just retired to deliberate a verdict on some of the ideas we present in this book.

But our brief foray into physics envy introduces another, more general, sort of a myth: the various "isms" which purport to tell us how we ought to look at *Homo sapiens* and the place our species occupies in nature. If the proper study of mankind is man, we cannot expect to get too far if we want to know to what extent our evolution—physical evolution, behavioral evolution—resembles the histories of nonhuman organisms, particularly those great apes which seem so (wonderfully, distressingly) similar to ourselves. On the other hand, if we accept the premise of some of our more vocal, if uninformed, pundits—that it's "all in our genes"—it seems we must look entirely elsewhere (baboon troops are an especial favorite) to find out about who we are, why we do what we do, and how all this came to be. Reductionism and its strict converse (in this instance, the notion that the only thing interesting about mankind is our set of unique properties) are just the most general of the various formulations of how we should go about understanding ourselves and how we got here. The notion that mankind is to be understood as a "naked ape" is about as useful and informative a springboard for a research program as the notion that we were divinely created in God's very image. Both views are mythic; and each has its modern, relatively sophisticated manifestations.

In arguing against either extreme we are not merely calling for application of the golden mean to solve a dispute. We are as passionately wedded to our own ideas as is anyone else to theirs. But the man as an animal vs. man as an angel dichotomy overlooks the real task: our physical and cultural evolution involves many components, each with its own distribution in the natural world. We share some of our features (most of them physical, some behavioral) with other creatures, while some others (some physical, most behavioral) are unique to ourselves. The problem is to find how widely each property is in fact distributed. That this is, in some instances, no easy task is not to deny that it is the best way to tackle the problem of who we are and how we came to be here.

CHAPTER TWO

Bestial Origins, Godlike Aspirations: Mythic Views of Man's Place in Nature

Few of us would deny that ours is a species of culture-bearing animal. This is the commonsense view of our nature. We in the Western world have thought of ourselves as an essentially peculiar animal, but an animal nonetheless, at least since the days of the ancient Greeks. Demosthenes, albeit jokingly, defined man as a "featherless biped." Aristotle classified us among the animals. And we find general acceptance of our status as animals on up through history: the Tunisian sage Ibn Khaldun, writing in the fourteenth century, said that "man, who shares with the other animals such animal traits as sensation, movement, and the need for food and shelter, is distinguished from them by his ability to think." It is difficult to say it better. Khaldun's view of mankind is so starkly simple and sensible, it is hard to imagine anyone's quibbling with it.

But quibbles there are. Two sorts of myths depart rather drastically from the balanced view of the Tunisian scholar. One myth—by far the earlier of the two—goes as follows: man, whatever his resemblance to animals may be, is something special, something above the animal realm. His nature, including his origins, are not to be understood in biological terms, particularly in terms of biological evolution.

This position comes in many guises. Societies, from the earliest records right on down to present times, have tended to look upon themselves as "the people," and all other societies of *Homo sapiens* known to them are "troglodytes," "trolls," or even "animals." In other words, there is a tendency to distinguish between one's own group (the "real" people) and other groups (the "not-real" people). Not-real people are ranked distinctly below real people. They are allied with animals, whereas real people are, correspondingly, not animals. Under this theme, eventually to grant other societies or races truly human status exalts them, removes them from the bestial ranks. Allowing human status to other societies never implies that one's own status is degraded down to the animal level previously attributed to the others. Lest anyone suppose that these attitudes died long ago or are at most the peculiar views of a few scattered, primitive peoples, one need only think of the attitudes of modern soldiers in combat. Denial of human status to one's opponent on the battlefield (as reported, for instance, in the Vietnam War) is an all too understandable mechanism to rationalize the killing. More subtly, denial of human status, to the point of characterizing members of other groups in unflattering animal imagery, lies at the heart of the expression of racist attitudes.

The more positive side of this myth stresses the "man as more than mere animal" point of view. The association between mankind and supernatural powers permeates religion, and the notion that God created man in his own image is an old one indeed. As the rallying point for morality and regularity in the conduct of social affairs, as an appeal to be civilized, to rise above base animal propensities, the notion of man's affinity to God has received a great deal of play, if with mixed success. When providing the basis of a moral ethic, the metaphor is both appropriate and appealing, as witness its extreme longevity. But a problem arises: to declare that man is not a creation of God in His image but instead a product, however exalted, of the rather more prosaic process of evolution seems to threaten the ethic—at least in the minds of the adherents—of some religious sects. Creationism, associated in the United States primarily with fundamentalist Protestant Christianity, asserts the literal biblical account of God's creation of mankind and the rest of the universe. Evolution, which says that whatever else man might be he is an animal, descended from other animals less Godlike than he, is anathema to creationists, simply because it seems to threaten the ethic implicit in the story of divine creation in God's own image. That there are other, more compelling reasons to retain a moral ethic within society eludes creationists. Today, creationism is once again on the rise, as part of a revival of populism in the United States and elsewhere.

But there are other versions of the myth that man's resemblance to animals is somehow unimportant to our understanding of ourselves. During the Renaissance, the Church's assurance of the godlike nature of man set the tone for Alexander Pope's injunction that the proper study of mankind is man. Though Pope was telling us to study man instead of seeking to understand the ways of God, his aphorism comes down to us as an assurance that we need look no farther than ourselves to understand our own nature. Surveying our own works, our ideas and aspirations, has preoccupied great minds for centuries. This pursuit, too, is appropriate. But its extreme incarnation—that somehow our arts and letters, our lofty ideas, and perhaps our views of the structure of economic systems and other social elements are all we need to understand the nature of man and whence he came—seems rather lopsided. All these works of man, and more, are certainly fit items to ponder, but is that all there is to us? A narrowness creeps in with this point of view, a myopia which can distort the very nature of human existence. Until comparatively recently, students at a major eastern university were inculcated with a view of the emergence of "modern man"—a creature who appeared in the Renaissance somehow different (and more advanced) than he was in medieval and earlier times. Such a view utterly collides with the distinctly human traits Björn Kurtén sees in his scenario of life in the Scandinavia of 35,000 years ago, in which two separate species of humans (Neanderthals and our own direct forebears) confront one another. Though a work of fiction, Kurtén's novel *Dance of the Tiger* imagines in a profoundly convincing manner the distinctly human qualities of both Neanderthals and ourselves. The stage on which the drama unfolds is an accurate reconstruction of the world 35,000 years ago—a mere drop in the bucket of time, given the several-million-year existence of our distinctly hominid lineage, but an immense span of time compared with the scant 500 years since the Renaissance. Focusing exclusively on our own recent (and Western world) products confines our vision of ourselves to an egotistically large but realistically narrow scope. We need something more.

What else is there? Well, let's look at ourselves as animals. Of more recent vintage, largely because of social (especially religious) opposition, the notion of man as an animal is now all the rage. Eureka! What a discovery!—though, as we have seen, the sober view that man is an animal, of however peculiar a kind, can be traced well back into written history, despite its relative unpopularity.

As a socially as well as scientifically acceptable idea, however, evolution is not very old. Barely more than a hundred years separate us from Darwin's classic *On the Origin of Species*, the document that tipped the scales, at long last, in favor of evolution. And it did not take too long for the message to be applied to humans, though Darwin himself at first merely ventured to suggest that his idea would "shed light" on the question of our own origins. "Evolutionism" quickly permeated our view of our physical origin and was also quickly incorporated into the fledgling science of anthropology. Cultural evolution, scenarios of the transformation of small bands of hunters and gatherers into chiefdoms and ultimately into the first states, became a commonplace in the nineteenth century, especially through the pioneering efforts of Edward Tylor and Lewis Morgan. The twin themes of biological and cultural evolution of mankind have been developed strongly in the past seventy five years. Yet, in spite of all this, the "news" that our species has a place in the animal kingdom seems to be front-page, best-seller stuff these days. There are, of course, various manifestations of the attitude that our knowledge of ourselves is to be construed, fundamentally, in biologic terms.

Best known are the works of Robert Ardrey (*African Genesis*), Elaine Morgan (*The Descent of Woman*), and others of this genre. Sensationalism appeals, to the point at which it seems that the American public has an insatiable apetite for anything developed by persons *outside* the ranks of scientific academe. Ardrey, Morgan, and their ilk interest us most at this juncture because they seek to explain us to ourselves in simplistic biological terms: we are who we are because our ancestors were who *they* were. But it is worth saying, in passing, that the same sensationalist appeal extends to other realms, such as the physical history of the earth (e.g., Immanuel Velikovsky) and the particularly nasty views of Erich von Daniken. *Chariots of the Gods* is based on the conceit that no earlier peoples, more "primitive" than ourselves, could possibly have built Mayan temples, Nazca ground lineations, Easter Island statuary, and the like. Therefore it *must* have been visitors from outer space. Such stories, plus fantasies about the Bermuda Triangle and UFOs, evidently have an inherent appeal far beyond the prosaic, usually rather commonsensical explanations preferred by reasonable people.

We are who we were. This is the message of the pop biologists who would have us understand ourselves in evolutionary terms. Concocting the story, based on carefully culled and rather scanty evidence, Ardrey, for example, creates the fiction that four-foot-tall *Australopithecus africanus*, roaming the African savannahs 3 million to 4 million years ago, was a vicious predator, a sort of "killer ape." We owe *our* viciousness, it seems, to this remote ancestor. It bears repeating that this is a cop-out—blaming our wars and violent crimes on some remote ancestor rather than placing blame squarely where it belongs: on the shoulders of *Homo sapiens*, "man the wise."

Not all current attempts to focus explorations of human nature and origins within the sphere of biology come from pop writers. We now have sociobiology, an academic discipline largely, if not wholly, biological in nature, which seeks a universal "science of society." Human beings, it is agreed, are not the only social creatures in nature. Some ants and bees have social organizations, many of them quite complex. Behaviorally minded biologists have labored long to understand the mechanics of such nonhuman societies and how they might have evolved. Perhaps it is not surprising that they have started openly to speculate about the implications of their research for understanding patterns of human social organization.

Looking for elements common to all societies in nature, sociobiologists have attempted to ascribe a biological basis for various elements of the behavior of humans, both as individuals and in groups. There is nothing seriously wrong with such an approach. Some aspects of our behavior are distributed far beyond our own species and even beyond the confines of our own primate branch of the mammals. Sexual reproduction, which is an aspect as much of behavior as it is of physiology, is as old as the earliest true animals—at least 700 *million* years old. There are general patterns of mammalian sexual reproductive behavior geared to internal fertilization and implantation of the fertilized ovum in the wall of the uterus. Primate (i.e., lemurs, monkeys, apes, and men) sexual behavior differs in detail from patterns seen in, say, the dog and cat families. And, as a general law of nature, every species, certainly including our own, has a particular behavior unique to itself to insure mating with members of one's own species. There is nothing intrinsically wrong with looking for a remote biological basis for particular elements of human behavior.

Yet sociobiology has become, not altogether unfairly, synonymous with a highly reductionist approach to understanding human behavior. A general "science of society" seeks to root out the ingredients common to all societies. A certain amount of altruism, for example, where some individuals appear to act contrary to their own apparent best interests in order to serve some "higher" purpose for the good of the group as a whole, seems to be a necessary ingredient of social organization. Such observations are instructive from a purely functional point of view: evidently no society, as the word is conventionally understood, can exist without those features observed in all. But what does this tell us about the origins of specific societies?

To answer this question, we will propose an analogy. Sharks and porpoises, plus the extinct ichthyosaurs, share a highly distinctive general body plan. As highly mobile, efficient predators, these marine creatures have streamlined bodies with very similar sets of fins. Conventional wisdom (and we see no reason to demur) has it that this particular body design represents an ideal engineering solution to the problem of high mobility of active predators living in a watery environment.

Though they were classified together by some early biologists, it has long been known that sharks are rather primitive vertebrates, that porpoises are whale relatives, thus mammals, and that ichthyosaurs are related to dinosaurs and other land-living reptilian creatures of the Mesozoic era. There can be no "science" of the group "sharks + porpoises + ichthyosaurs" because such a group never existed. All we learn, by studying them collectively, are a few principles of hydrodynamics as applied to a particular engineering problem: designing a body to fit a particular ecological niche.

Still, evolutionary biology has been profoundly hung up over the notion of adaptation. To this very day, adaptation by means of natural selection is viewed as *the* fundamental process of evolution. Biologists have been so entranced with the idea of design in nature, the modification of structure and behavior to perform a particular function, that they have condoned studying the "evolution" of such nonexistent groups as "sharks + porpoises + ichthyosaurs."

But such arguments make no sense. Evolution is, first and foremost, a matter of ancestry and descent. The groups that evolutionists really should seek to study (and are now returning to) are natural groups of species all descended from a common ancestor. Yes, some insects have societies, and yes, so do we. But bees are more closely related to other animals (for example, solitary wasps) than they are to us. "Social animals" is not a natural group created by evolution. All we can gain from lumping them all together is the analog of the engineering principles gleaned from "sharks + porpoises + ichthyosaurs": interesting, but of highly limited application. In their zest for universal generalizations, sociobiologists have violated the cardinal rule of evolutionary biology: explanations of evolutionary history must be made for groups that evolution has produced.

But it *is* appropriate for anthropologists and biologists to compare elements of the social organization and other aspects of human behavior with other, closely related organisms. All primates possess social organizations of one kind or another. Here is a group—Order Primates—produced by evolution. Here is where we should actively seek out the distributions of behavioral traits to understand better when, in the course of evolutionary history, each of our own traits first appeared.

Thus we have a Scylla and Charybdis—a dangerously narrow passage between the mistake of total reduction to vaunted (and frequently already outmoded) biological principles on the one side and the notion that man stands alone, apart from nature, on the other. Viewed this way, we certainly have a problem.

If we return to the good Ibn Khaldun's view, however, the problem vanishes. The answer to both sides is "a plague on both your houses." We need only to keep clearheaded. We admit that our species, *Homo sapiens*, is an animal: Genus *Homo*, Family Hominidae, Superfamily Hominoidea, Order Primates, Class Mammalia, Phylum Chordata, Kingdom Animalia. We are nested in ever larger boxes, which include ever more animals, as we ascend the classificatory hierarchy which summarizes our evolutionary relationships in nature. We can now take each trait—anatomical, biochemical, physiological, behavioral—and see how widely it is found, starting with our own species and working out: other species of the genus *Homo* (now all extinct), other Hominidae (our remaining fossil relatives), other Hominoidea (including the greater and lesser apes), and so forth. This is a mapping exercise, and it is not always easy to perform. But it does tend to cut through the mounds of tendentious rhetoric given us by the spokesmen for Scylla and Charybdis.

It is in this way, by analyzing the distributions of characters—specifically, evolutionary novelties—among living things that we are able to reconstruct the ever more inclusive hierarchy of which we ourselves form part. And seeing ourselves in this wider context, the context of life as a whole, gives us a salutary perspective upon our place in the natural world. For example, it has lately become fashionable to express amazement at the "remarkable" genetic similarity—amounting perhaps to 98 or 99 percent of genes—between us and our closest relatives, the great apes. Maybe this stems from reluctance on the part of an egocentric species to see those things that separate it from other living beings as the relatively minor distinctions that they are. Whatever the cause, we tend to look at the genetic similarities between ourselves and the apes as being surprisingly small given the manifold differences, anatomical and behavioral, that we perceive. But in doing so we only deceive ourselves. If we look at the vast array of living forms on earth—algae, trees, clams, spiders, ourselves—what we should surely be astonished by is the fact that the genetic differences between us and the chimpanzee is as *much* as 1 or 2 percent. The differences between humans and chimps fade into utter insignificance when we compare either with a tree. But all three organisms, as a legacy of their remote common ancestor, possess eukaryotic cells, enclosed in a double membrane and containing a well-defined nucleus. And between the two hominoids and the tree lies a vast array of living things. Rather than egotistically to marvel at the magnitude of the objectively tiny differences between us and chimpanzees, we would do far better to try to understand what these differences mean in the almost infinite variety of the natural world. We can fully comprehend this significance only by realizing that like any other living species we are a mixture of characteristics held in common with others, and of properties unique to ourselves.

CHAPTER THREE

Evolution: The Myth of Constant Adaptive Change

WHEN ASKED for a one-word definition of evolution, most people say "change." The very word connotes movement, alteration, progression—change. Perhaps even inevitable, inexorable change. When they are pressed to be more specific, the single-word definition of evolution most people give is "development." We speak of the evolution of the universe, of the solar system, of the earth, and of course, of life. But we also speak of the evolution of language, of culture, of economic systems, and even of ideas. In many languages, the word for "development" may also mean "evolution": *Entwicklung,* a German word, is an example. Usually meaning development, as from an egg to an adult, entwicklung can also refer to the evolutionary history, or development, of entire groups of organisms.

Thus, the kind of change most people have in mind when they use the word "evolution" is not a haphazard, anything goes sort of affair, but a much more definite alteration of state which follows a regular and understandable, if not entirely preordained, course. The sort of change envisioned is a logical development. It proceeds from the simple to the complex, from the primitive to the advanced, from the imperfectly formed to the perfect. Evolution above all connotes progressive improvement.

But to an individual or society by which stability is valued and fixity of all natural things is the predominant world view, change is anathema. Evolution in the sense of change is a threat, a frightening prospect to be resisted. Most societies collectively adopt this point of view most of the time. People are notoriously resistant to change, real or imagined.

Occasionally, however, societies will embrace a concept of change—so long as it is of a progressive, improvement-oriented kind. At such times the social philosophers, if not the populace as a whole, will grant the possibility of modification and alteration as long as it is positive. It was in just this sort of intellectual climate that notions of evolution—of the universe, of life, and of mankind both physically and culturally—caught fire. Fueled by visions of ever-expanding economic wealth and perhaps fanned by the whirlwind of rapid technological change begun by the Industrial Revolution, the notion of progress came to dominate the world view of Western social theorists during the nineteenth century.

Though he led a rather reclusive life at his house at Down, in Sussex, after his five-year trip around the world in the *Beagle*, Charles Darwin was no isolated individual who managed somehow to concoct a brilliant generalization about life. Darwin was very much a product of his times. He read widely and had a voluminous correspondence. He was keenly aware of the notions of change in the air—notions challenging the ages-old view of the fixity of species in biology, and similar ideas of change and progress among economists and other social theorists of his day. Darwin did not invent the idea of evolution: he made instead a compelling case for it in biology. And he did so in the context of his time, drawing the essence of his notion of biologic evolution from ideas current in contemporary society and arguing for the validity of his views in terms others could understand.

To Darwin, evolution was "descent with modification." He saw a pattern in nature, a hierarchy of similarity linking all forms of life, a pattern seen by Aristotle and other ancient Greeks and the subject of serious biological research since Linnaeus established his scheme of classification a century earlier. Darwin saw that the simplest explanation for this pattern of degrees of resemblance among organisms was the simple notion that they were all related. The more closely similar two animals or plants may be, the more closely related they must be. In his vision, Darwin saw all organisms descended from a single common ancestor in the remote past. And just as family histories can be drawn on a piece of paper, the genealogy of all life could be depicted as a branching diagram—a tree.

The simplicity of this idea—that all organisms are related—is startling. Instead of hundreds upon thousands of mystical separate acts of creation, the entire diverse array of life could be explained as a pattern of ancestry and descent from an initial single-step beginning. But to convince the world—other biologists but also the rest of the scientific community and society in general—of the validity of this grand, sweeping notion, Darwin had to eliminate the notion of fixity of species. Old species, in his evolutionary concept, had to give rise to new ones. The old notion that species, once formed, were forever to remain the same, had to be replaced by a new notion which saw species as fluid, changeable entities. Species might *seem* to be stable entities, quite distinct from each other in the world around us. After all, little confusion among the bird species in the gardens of Europe is possible; they seem on the whole quite different and distinct from one another. But Darwin argued that given enough time, their apparent fixity melts away. The distinctive features of two wren species, should we be able to trace their pedigrees far enough back into the recesses of time, would gradually come closer and closer until we would find them melded together in a single common ancestral species somewhere in the geological past.

Gradual, progressive change: this is the hallmark of the Darwinian concept of evolution. In its basic form, it is no different from other concepts of change popular in Victorian England. Darwin got his basic notions of how evolution works from general ideas of change around him, and he characterized evolution in these terms as a technique to convince his peers that species are not the indelibly fixed sorts of entities biologists had long considered them to be. If change were to be admitted, if life were to be granted a history, then Darwin not only preferred a vision of gradual change but needed to argue in such a fashion to convince his peers. A truly vast period of time, over which small, imperceptible changes could gradually accumulate, seemed the best way to attack the notion of species fixity. The gradual, progressive accumulation of small changes was an idea much more in tune with prevalent notions of progress in social change, for, while the notion of change had taken hold as an explanation for post–Industrial Revolution events in society, it was wedded strictly to the notion of progress. Things were getting better. It was the nature of things for economic systems and societies in general to *improve*. Change is inevitable; it is progressive improvement. And, seen as progressive improvement, change is slow and steady. It is gradual. Improvement comes after trial and error weeds out the old and less efficient, keeping the newer, more efficient—whether the change be in marketing techniques, banking systems, manufacturing mechanisms, or anything else. In Darwin's world, sudden, radical change was as abhorrent as ever (though Karl Marx was writing in the library of the British Museum at that very time). Whether Darwin's concept of biological evolution was so similar to general notions of progress within society at large because he was unable to see it any other way or because he consciously used it as a ploy to convince his peers is moot. The point is, his concept of evolution—a view still very much dominant today—is distinctly Victorian.

The similarity between Darwin's portrait of the evolutionary process and general Victorian notions of progress goes still deeper. Darwin spent much of his epic monograph *On the Origin of Species* (1859) establishing the fact of evolution and arguing for a plausible mechanism, an explanation of how evolution actually works. He had to argue that species are not fixed and he had to establish a causal process which could create the patterns of change through ancestry and descent. Again borrowing ideas outside his field, he read the economists Thomas

Malthus and Adam Smith and from them discovered the notion of competition for resources. More organisms are produced each generation than can possibly survive. Darwin calculated that a single pair of slow-breeding elephants, producing an average of 1 offspring every 10 years, would leave a population of 15 million descendants in only 500 years. Obviously, in the economy of nature, there is a limit to growth. Resources for all species are finite, setting the limits for population growth.

In the competition for resources, there are winners and losers. The winners survive and produce offspring. The losers simply lose. The best survive, and since their offspring resemble them, there is improvement of the population as a whole as time goes by. This is the basis of Adam Smith's laissez-faire, open-competition economics. It is the basis of ideas of progressive social change in the theories of such Victorians as Herbert Spencer (who coined the phrase "survival of the fittest"). And it is the prevailing notion of natural selection.

Natural selection, in Darwin's time as well as today, is nothing more than differential reproductive success. In a world of finite resources, more offspring are produced than can survive. Those best suited to the environment survive and reproduce. The notion embodies a simple mechanism for progress. Only the best are "selected" each generation to produce the next generation. As new variations appear, the opportunity for ever greater improvement follows. Darwin knew nothing of genetics as we understand it today, but he did realize that organisms vary within populations, that offspring tend to resemble their parents, and that new characteristics occasionally appear unexpectedly in some progeny—the only three items concerning inheritance necessary for the theory of natural selection. Thus, a gradual, progressive improvement is to be expected within a reproductive community, even if the environment remains the same for millennia. Selection is constantly working to improve the stock.

If natural selection can improve the adaptation of a species, that is, the behavioral and anatomical characteristics it has developed for its particular mode of life, the idea of change in response to a changing environment becomes all the more plausible. As conditions change (so goes the idea), organisms once best suited to the environment find themselves less well adapted. Selection works to promote other variants which are the best suited to the new environment. Any long-term change in environmental conditions would inevitably result in a long series of progressive changes, as natural selection tries to keep the species up to match the changing times.

Thus, change is viewed as inevitable. It will be slow and gradually progressive in a constant environment. It will be faster and more marked as the environment changes. From this perspective, fixity of species becomes a logical impossibility. Species *cannot* remain substantially the same through time. Change is inevitable, albeit gradual change by small increments, the sort of change only detectable by examining long sequences of fossils collected carefully through a thick expanse of rock. This is the Victorian notion of progressive change in general, and Darwin's view of the evolutionary process in particular. And, until recently, it has been the predominant view of modern evolutionary biology.

Modern Evolutionary Theory

Darwin's characterization of biological evolution as "descent with modification" deserves a closer look. This deceptively simple phrase actually embodies two different notions: change itself, plus a concept of genealogy. The two ideas are closely related but are certainly not identical.

It is the "descent" part of Darwin's phrase that has caused the most trouble to evolutionists in the 120 years or so since the phrase was penned. "Descent" has proven to be tricky mostly because it has largely been ignored as a serious question. To be sure, debate rages on, sometimes even on the front pages of newspapers, about ancestors and "missing links"—usually provoked by new discoveries of fossils belonging somewhere in our own human pedigree. And, of course, paleontologists and biologists continue their efforts to classify the elements of life, a task which presupposes an accurate notion of the actual history of life. The practical work of reconstructing life's genealogical history has hardly been ignored. But the theorists—those who would explain *how* life has evolved—have fallen down on the job by not examining the concept of "descent" as carefully as they have that of "modification."

Darwin's view of natural selection, which biology retains in essentially similar form today, saw change occurring from generation to generation. It would be logical to conclude, then, that the ancestors in evolution are fathers and mothers, and the descendants their sons and daughters. But the differences between great-grandparents and great-grandchildren, however great or small they may seem, are not the stuff of evolution. This is because, in sexually reproducing organisms such as ourselves, there is a mixing of pedigree in each successive generation. So Darwin's view—and that of modern population genetics as well—focuses on the statistical frequency of features changing in a breeding group from generation to generation. (Darwin spoke of anatomical and behavioral characteristics. Modern evolutionists do too, but geneticists have, in addition, developed a theory of generation-by-generation change in gene frequencies, which underlie the observed anatomical and behavioral modifications.) The ongoing system of parenthood produces the pattern of generation-by-generation change. In a very real sense, generations within an entire reproductive community form the ancestors and descendants of evolution.

Yet, Darwin's main title for his most famous book was *On the Origin of Species*. A large segment of contemporary evolutionary literature is devoted to the problem of how new species arise from old ones. This is just another way of saying that some species are ancestral to other, descendant species. Perhaps it is *species* that are the ancestral-descendant units of evolution.

Darwin never addressed the question of the origin of species in his book. The reason is simple: despite his title, he didn't see it as a problem. After all, he was intent on destroying the notion of fixity of species. If species are not forever immutable, if instead they are ever-changing, then no matter how distinct one from another they may appear at the present time, they were closer to each other in the past and presumably will be even more different from one another in the future. Thus, species to Darwin were nothing more than the lowest category of organisms on Linnaeus' hierarchical totem pole. Species are reproductive communities recognized as species because their members form a parental network with each other and not with members outside the community—and because as a result of this all the members within the community look more or less alike and tend to look different from members of other reproductive communities. But for Darwin's notion of evolution (*change*) to work, these communities could not be stable. Those very anatomical and behavioral properties which stamp them as unique were different in the not-too-distant past and are bound to be different in the near future. And though Darwin himself did not specifically say so, biologists treading in his footsteps have long been fond of pointing out ("speculating" is perhaps a better word) that if enough generations go by, reproduction would be impossible between earlier and later members of these reproductive communities. In other words, a time-machine experiment in which some of us traveled back into the Ice Age and stopped from time to time to mate with ancestral members of our lineage would eventually reach a point of repeated failure to breed because of the intrinsic differences between ourselves and our remoter ancestors. And the equally fanciful vision of traveling into the future would produce entirely similar results.

These twin notions—that the passage of time inevitably leads to physical and behavioral changes and that the reproductive network binding a species together is somehow destroyed in the process—imply that species are not actually "real." Instead, they are ephemeral breeding communities which inevitably and inexorably transform themselves out of existence. This is how Darwin saw them. This is how we still tend to see them today.

This pervasive view of life has led to a habit among paleontologists of saying that one species has evolved from another when really what is meant is that so much change has accumulated in a lineage (a reproductive community unbroken through many generations and thousands, or perhaps millions, of years) that we should give the later member a new name. In other words, while we can be pretty objective in recognizing discrete breeding communities (species) in the modern world, the picture of species in time (as seen in the fossil record) is one of such constant change that naming species is an entirely arbitrary matter of chopping up an evolutionary continuum. The "origin of species" is thus an artifact—the product of the paleontologist classifying fossils—rather than an actual phenomenon of nature. We must name organisms to talk about them, so we chop these evolving lineages up as best we can. Gaps in the fossil record, chunks of geologic time for which no fossils belonging to a particular lineage have yet been found, are often cited as helpful to those who are faced with the task of dividing up these evolutionary continuua. Biologists following this line of thought to its logical conclusion are fond of expressing thanks that the fossil record is not complete. Not only do the gaps provide the convenient points for the definition and naming of species, but even more important, if the fossil record were complete, we would have to classify all life into a single species unless we were to be utterly and capriciously arbitrary. However, the very fact that a theoretical notion has led serious scientists to state publicly that they are glad their data are less than perfect is a sure sign that something must be wrong with the ideas themselves.

The Synthetic Theory of Evolution

The prevailing view of the evolutionary process is called the "synthesis" because it integrated, in the 1930s and 1940s, the seemingly disparate data of genetics, systematics (study of species and their classification), and paleontology into a single, coherent theory. Though it seems obvious that if life evolved, the facts and ideas of genetics and paleontology which seem so different, nevertheless *must* fit together, a generally agreed-upon integration of these disciplines proved elusive for the first seventy years or so following the appearance of *On the Origin of Species*.

The synthesis arose only after some initial controversies in the relatively young science of genetics had been resolved. Darwin, as ignorant as his peers about the mechanisms of inheritance, was able to treat the subject as a "black box." His precise views on inheritance, while wrong, did not affect his theory of descent with modification, driven by natural selection. But once the fledgling science of heredity got going around the turn of the century (with the rediscovery of the Austrian monk Gregor Mendel's work), new ideas in genetics had to be reconciled with Darwin's notions. Particularly troublesome was the notion of large effects: Gregor Mendel's peas were either wrinkled or smooth, yellow or green. A mechanism for inheritance based on discrete factors, "genes," was indicated. The either/or state of the characteristics studied by early geneticists seemed hard to reconcile with the Darwinian assumption of blending inheritance necessary for a smoothly gradational mode of change from generation to generation.

Worse yet was the notion of "mutation" developed by the Dutch botanist Hugo De Vries. Studying evening primroses, De Vries made much of the "sports"—changes appearing in individual flowers on plants—that would arise from time to time. These large-scale changes between parent and offspring plants must mean that something new had occurred, a change in the genetic material itself. De Vries elaborated these observations into an evolutionary theory that saw change as sudden and abrupt—a far cry from Darwin's much more moderate claims.

Thirty years of intense experimentation and theoretical analysis saw a reconciliation of sorts between the expected small-step changes and the challenges brought by notions of particulate inheritance and large-scale mutation. The newly developed genetic theory held that the "one gene-one character" notion was simplistic. Few features are inherited as simply as the wrinkling of Mendel's pea skins. Most features are developed by more than one gene, and each gene is involved in the development of more than one character.

Work with mutant forms, furthermore, showed that De Vries' mutations (now known to involve large changes of chromosomes) were on the high side of a spectrum of mutations. Most mutations have far smaller effects, according to the genetics of the 1930s. This view is still essentially held today, though in the 1930s the chemical anatomy of inheritance (the structure and properties of DNA) were virtually unknown. Thus, the way was paved to reconcile the theory of inheritance with the older view Darwin espoused that changes from generation to generation within species would be gradual. Then population genetics, a branch of theoretical genetics, took these notions and developed a comprehensive mathematical theory showing how frequencies of alternate forms of genes (alleles) could change within populations through time, given certain mutation rates and intensities of natural selection. By the thirties, genetics had been reconciled with Darwin's outlook. Evolutionary theory, essentially stagnant while geneticists worked out their early problems, was now ready to receive another look.

The essence of the view that emerged with this new look is deceptively simple: natural selection, working on a groundmass of genetic variation, changes gene frequencies each generation. Mutations are the ultimate source of the variation, but it is selection, working to perfect adaptations or to keep a population in step with changing times, that is the real agent of generation-by-generation change. Over long periods of time— thousands, millions of years—such minute step-by-step change will have large effects. So large, in fact, that all the evolutionary patterns seen by systematists working on living plants and animals, as well as all the fossils seen by paleontologists, are nothing more than the results of these small-scale processes summed up over geologic time.

This is an exciting idea: that one fundamentally simple process different only in detail from Darwin's original notion could account for the entire evolutionary history of life. The logical requirement that any theory of the evolutionary process must, after all, account for all aspects of the evolutionary history of life was solved by reducing the phenomenon to its barest essentials: the nitty-gritty of change in frequencies of genes within populations generation by generation. Now paleontologists could relax. True, it seemed more than ever that the fossil record was too poor to study the process of gene frequency change. But that no longer mattered; even were the fossil record complete, the mechanics of evolution would be accessible only to the geneticists in any case. And, not surprisingly, after a masterful demonstration by George Gaylord Simpson that the data of paleontology are in fact consistent with these views, paleontologists have been as silent as the rocks they search on the subject of evolutionary theory. They had been removed from the game. The geneticist studies the mechanisms of evolution. Systematists and paleontologists study the results. All a paleontologist need do is extrapolate the findings of genetics to ask how the generation-by-generation process of neo-Darwinism looks in geologic time.

The Adaptive Landscape

In 1932, the geneticist Sewall Wright developed a simple yet effective imagery to convey some of his ideas about relative "success" of genes in breeding populations. Each gene is a locus, or place, on a chromosome. Each gene has one or more forms, or "alleles." With thousands of loci, each with several alleles, one would expect some of the many possible combinations of these alleles to be more "harmonious" than others, as Wright put it. The more harmonious combinations are those that produce the fitter individuals—the ones better equipped to flourish in the environment. But Wright posed a problem and to dramatize it he drew a crude map. Hills and valleys were delineated by contour lines. On conventional topographic maps, the contour lines connect areas of equal elevation. On Wright's concept of the "adaptive landscape," the contour lines connected hypothetical regions of fitness. On the tops of the hills sat the more harmonious allelic combinations; in the valleys reposed the less fit individuals. The problem, as Wright saw it, was this: how does a species maximize the number of individuals sitting on the peaks?

The power of this imagery is compelling. Others soon picked it up and used it for purposes far beyond its original intent. Today, no textbook on evolution is without this pictorial metaphor—and seldom is the difference between Wright's original use and the more typical notions developed in the 1940s pointed out.

The adaptive landscape in its familiar guise sees the hills and valleys as environments to which populations or species of organisms are adapted. The hills are ecological niches to some and simply peaks of adaptive perfection of the occupant population to others. The valleys are inhospitable areas unoccupied for any length of time: literally, valleys of the shadow of death. The difference between this image and Wright's original metaphorical geography is twofold: entire populations or species occupy the hillsides, and there are environmental differences between adjacent peaks. Each peak is a different ecological niche. The problem then becomes: how do species cross the valleys and climb the next peak?

As it was developed in the 1940s, the metaphor of the adaptive landscape is used to explain how adaptation occurs via natural selection. The entire history of life falls out from this picture of changing landscapes followed by adapting species, and this explains why the adaptive landscape is the favorite graphical image of the synthetic theory of evolution.

Though some biologists did, indeed, theorize how a species might abandon one peak, traverse a valley, and climb another peak, by far the greatest use of the landscape model focuses on modifications in the map itself. As time goes by, the positions of the hills and valleys gradually change, "more like a choppy sea than a static landscape" (in the words of the paleontologist G. G. Simpson). Selection is constantly at work to keep track of these changes, to make sure the hilltop does not move out from under the species perched on its top. It is change or die. Selection will act to perfect adaptation to the peaks while the landscape is stable (though it is always good to "stay loose," not throwing all your adaptive eggs in too narrow a basket in the very anticipation of inevitable changes in the landscape). But the landscape is more changeable than static. Over time the landscape will inevitably change. So directional natural selection keeps a species running, effectively playing catch-up football with the fickle environment.

This imagery is exactly the same as the conventional view of directionally shifting gene frequencies through time, and as such is no radical departure from conventional neo-Darwinian ideas. But in it lie the seeds of a truly synthetic explanation for the development of life's diversity. In the 1930s and 1940s, the figure for the total number of living species was usually given as 1 million. (Estimates of 10 million or more are common today). Natural selection, doggedly tracking environmental change, could theoretically modify a species ad infinitum: but, if all life is descended from a single ancestor, how did we get so many different species? Whence the diversity? Even though Darwin argued against the fixity of species in time, the fact that there is a number of readily separable kinds of organisms, breeding among themselves but not with other kinds, is indisputable evidence that separate lineages somehow become established in the evolutionary process. How?

The adaptive landscape explains all. Peaks on the landscape do not merely change position on the map—they sometimes divide into two peaks! As time goes by, the cleft between the two peaks becomes deeper, the distance between the tops greater. What was once one species occupying one peak gradually becomes two as the peak divides and remnants cling to each of the two new peaks as they move apart. The geneticist Theodosius Dobzhansky was especially enamored of this vision. He saw the peaks as niches: the reason why we have as many species as we do is that there are that many ecological niches. And niches can change; indeed, they can subdivide. Separation—the origin of new separate reproductive lineages (two or three where once there was one) is a matter of selection tracking subdivisions of peaks in the adaptive landscape. Speciation is nothing special; it is just another aspect of adaptation.

This hurdle past, the entire diversity of life falls into place. Over time, newly split-off peaks become far removed from one another, creating space, as it were, for numerous satellite peaks to form around them. What once were two closely related species in the Eocene, occupying adjacent peaks in the pictorial metaphor, continue to diverge and proliferate, until in Recent times, we look back and see, for example, the horse and rhinoceros families. Each has had a long history. And it is indeed true that the earliest horses and rhinos were quite a bit alike: far more similar to each other than are their descendants today. The imagery fits the facts rather well. And it gives us a single, unified, and charmingly simple explanation of the evolutionary process, an explanation which encompasses gene frequency changes from generation to generation, all the way up through the origin of different phyla and even the five currently recognized kingdoms of life. It shows how the evolution of kingdoms is in principle no different from the shifts in gene frequency seen within species. In fact, it says that all evolution reduces to nothing more than change in gene content and frequency within species. The process is adaptive, as natural selection guides change in response to changing environments. It is, at once, a simple view and a sophisticated version of Darwin's original conception. And it is, in a profound sense, probably wrong. It is one of the greatest myths of twentieth-century biology.

Evolution—Another Look

The conventional, synthetic view of evolution we have just characterized is profoundly reductionist: it reduces all evolutionary phenomena to a few simple statements about mechanisms. All evolutionary patterns in nature are produced by the same elemental processes. What is wrong with the synthesis is not the core neo-Darwinian formulations of mechanics (natural selection working on variation within species to effect gradual change). What is wrong is the wholesale, uncritical—and unwarranted—*extrapolation* of these mechanisms via metaphors such as the adaptive landscape to embrace the evolution and diversification of all life. In the last analysis, what is wrong is the myth that species are concrete realities at one time but when viewed through time lose any semblance of discreteness and somehow become less "real."

As we have seen, Darwin effectively had to deny that species are "fixed entities" just to establish the notion of descent with modification. But naturalists had to confront the simple fact that species are discrete entities in any local area. Evolutionists had to confront the persistence of taxonomists who continued to describe species as if they were something more than a mere arbitrarily assembled collection of similar specimens. At the very time that Dobzhansky and Simpson were developing their versions of the synthesis, the zoologist Ernst Mayr spoke of species as analogous to individuals of the microscopic organism *Paramecium:* even when one individual splits to form two, and in so doing complicates the notion of individuality, there was nonetheless one individual at the outset and two at the end. So it is, Mayr argued, with species: they are entities, things, and to that extent individuals. They occasionally split to form two or more new entities where once there was one. In the process of splitting, there may be some confusion about individuality, but when the dust settles, the individuality question is again itself settled. There are two or more individuals where once there was one. Species, at any one time, are real, discrete entities.

When it came to thinking of species through long periods of time, however, Mayr agreed with Simpson, Dobzhansky, and the other synthesists: by the very nature of the evolutionary process (meaning adaptation by natural selection—the very essence of the neo-Darwinian model), species must inevitably change. And in so doing, they must give up any sense of identity, discreteness, or individuality. Thus, in the long view, Mayr too adopted a reductionist standpoint, though his focus on species as discrete entities (later much more emphatically expressed) provided a dash of anomaly within the synthesis. For if species are real in time as well in space, the synthesis crumbles, and much of our conventional expectation of what the results of the evolutionary process ought to look like changes drastically.

What are Species?

The alternative view of species is an extension of the old naturalists' view: species are seen as real entities, reproductive communities separate from each other in space, but also discrete in time as well. They have beginnings, histories, and, inevitably, ends. Species have integrity, an individuality, throughout their existences—which may be as long as 5 million or 10 million years.

The main impetus for expanding the view that species are discrete at any one point in time, to embrace their entire histories, comes from the fossil record. Paleontologists just were not seeing the expected changes in their fossils as they pursued them up through the rock record. Instead, collections of nearly identical specimens, separated in some cases by 5 million years, suggested that the overwhelming majority of animal and plant species were tremendously conservative throughout their histories.

That individual kinds of fossils remain recognizably the same throughout the length of their occurrence in the fossil record had been known to paleontologists long before Darwin published his *Origin*. Darwin himself, troubled by the stubbornness of the fossil record in refusing to yield abundant examples of gradual change, devoted two chapters to the fossil record. To preserve his argument he was forced to assert that the fossil record was too incomplete, too full of gaps, to produce the expected patterns of change. He prophesied that future generations of paleontologists would fill in these gaps by diligent search and then his major thesis—that evolutionary change is gradual and progressive—would be vindicated. One hundred and twenty years of paleontological research later, it has become abundantly clear that the fossil record will not confirm this part of Darwin's predictions. Nor is the problem a miserably poor record. The fossil record simply shows that this prediction was wrong.

The observation that species are amazingly conservative and static entities throughout long periods of time has all the qualities of the emperor's new clothes: everyone knew it but preferred to ignore it. Paleontologists, faced with a recalcitrant record obstinately refusing to yield Darwin's predicted pattern, simply looked the other way. Rather than challenge well-entrenched evolutionary theory, paleontologists tacitly agreed with their zoological colleagues that the fossil record was too poor to do much with beyond supporting, in a general sort of way, the basic thesis that life had evolved. Only recently has a substantial number of paleontologists blown the whistle and started to look at the evolutionary implications of the marked pattern of *nonchange*—of stability—within species so dominant in the fossil record of life.

The coherent, static packages of anatomy found in the fossil record have always been called "species." If species are reproductive communities, though, how are they recognized in a fossil record where no direct evidence of reproduction could possibly exist? And given enough time (thousands and even millions of years) was it even likely that the older members of these "species" could interbreed with individuals from the later stages of the species' history?

Species in the modern world are indeed discrete reproductive communities. A network of parentage glues the community together and keeps it separate from other species. It is easy to visualize this pattern of parentage going on for thousands and thousands of generations—solving the second dilemma just raised: we need not worry if the older members of a species could mate successfully with later individuals, so long as there was an unbroken chain of reproduction linking them up through time. Reproduction between far-flung members of a modern species is sometimes impossible too, but the species remains an integral whole if the disjunct populations are linked by a chain of other populations whose members interbreed freely. Species cohesion in time as well as in space depends upon continuity of the reproductive network, not upon a hypothetical breeding experiment which tests the interfertility of two individuals from remote ends of the distribution of a species.

But without direct evidence of such reproductive continuity, how are we to call our groups of similar fossils "species"? The answer comes from the old naturalists' view of species fixity in the modern world—the simple observation that there are fundamental units, or kinds, that everyone calls "species," and which come in the form of separate, usually easily differentiated "packages." Individuals within species mostly all look pretty much alike and look rather different from members of other species, including their closest relatives. There are exceptions, such as closely related species which are confusingly similar but which nonetheless form separate reproductive communities. And some species are notoriously variable, spread out over a wide area of diverse habitats. But the overall picture is clear: the separate reproductive communities we call species are the same as the clusters of similar individuals that the old-time naturalists called species.

Paleontologists confront the fossil record knowing this. At any one spot in the rock layers, they see the same sorts of arrays of anatomical packages that zoologists and botanists—in fact, all of us—see in the world today. A paleontologist calls each of these packages a "species," knowing he might be fooled in some cases where two separate species are too similar to be told apart. Looking higher and lower, in younger and older rocks, he sees many of these same packages. And he sees in the vast majority of cases that these anatomical packages, these species, have remained substantially unchanged through monumentally long periods of time. Species, in other words, seem to be relatively static. There is frequently more variation throughout the geographic spread of a species at any one point in time than will be accrued through a span of 5 million or 10 million years.

This observation has two simple consequences, both of tremendous importance to evolutionary theory. First, Darwin's prediction of rampant, albeit gradual, change affecting all lineages through time is refuted. The record is there, and the record speaks for tremendous anatomical conservatism. Change in the manner Darwin expected is just not found in the fossil record.

The second simple consequence is the observation that species are stable and remain discrete, in time as well as space. They are individuals in the true sense of the word: they have beginnings, histories, and, ultimately, ends. During their life spans they may or may not give rise to one or more descendant species, just as humans may or may not produce children during the course of their lifetimes. To make a case for evolution, Darwin argued against the fixity, meaning reality, of species. Zoologists later returned to the notion that species are discrete, real entities in space, but continued to deny their individuality through time. It is now abundantly clear that species are real entities—individuals—in the fullest sense of the word. And it is these spatio-temporally discrete units which are the ancestors and descendants in evolution.

Speciation

New species arise from old. The prevailing view of the evolutionary synthesis saw speciation as no more than an accidental by-product of the *real* process of evolution: adaptation through selection, which effects so much change that we call the transformed products new species. Whenever two or more species appear where once there was but one, we invoke the imagery of the gradually subdividing adaptive peaks. Splitting—getting two species from a single ancestor—is merely a special case of the phenomenon of progressive adaptive change.

This view, favored by both paleontologists and geneticists in the modern synthesis, collides head on with the idea that species are discrete individuals. For the progressive, adaptive model of evolutionary change to work, species *must* be arbitrarily chopped-up segments of a continuum.

In a curious counterpoint to the notion of speciation through progressive adaptation, the very systematists working to restore a partial notion of discreteness and reality to species in the modern world were developing correspondingly rather different notions of speciation. Species are reproductively isolated communities. How does this reproductive isolation come about? And how does it relate to anatomical change in evolution? Species, as we have seen, are more than mere collections of similar animals or plants. Species have an identity and an integrity, an internal cohesion supplied by the process of reproduction. Nearly all organisms reproduce sexually, and the continued process of male-female mating creates a network of reproductive relationship which binds a species together. At the same time, the plexus of parental ancestry and descent keeps a species from blending with other species. Sexual reproduction is the key to the maintenance of the integrity and identity of all species. It is this reproductive aspect—far more than mere anatomical or behavioral similarities among its members—which lends a species its identity. Thus the evolution of species must involve, primarily, the establishment of new reproductive communities, rather than simple modification of anatomies and behaviors. It would seem that on these grounds alone the basic theme of the modern synthetic theory of evolution—adaptation by natural selection—does not address the fundamental problem of ancestry and descent in evolution: the origin of new species.

The synthetic theory treats reproductive isolation as a secondary by-product of the progressive adaptive divergence of an ancestral species as it follows environmental change into two new niches. As the ancestral niche divides, the species splits, and the adaptive modifications developed by each portion accumulate to the point at which the two are so different they can no longer interbreed.

But the data of the field naturalists point to a simpler explanation: the ability to reproduce successfully with members of another group seems to have rather little to do with degree of similarity in anatomy and behavior—the degree to which organisms share precisely the same adaptations to their environments. Some species, such as our own, are remarkably far-flung, developing great anatomical variability and cultural diversity. Yet the reproductive plexus is unbroken, and flying in the face of such diversity, still provides the glue holding the species together. Quite the reverse situation is also commonly found: many true species, fully reproductively isolated from their most closely related species, display hardly any anatomical or behavioral divergence from their ancestors.

Progressive adaptive change, then, cannot account for the sundering of species. Too many species, despite great variation (the result of adaptation to local environments) remain cohesive species, and too many true, reproductively isolated species have diverged but little from their ancestors. There simply is no hard and fast relationship between the origin of new species and the sorts of anatomical and behavioral changes which are the stuff of adaptive evolution.

Biologists have been aware of this state of affairs for years. From the 1930s onward, the hows and whys of reproductive isolation between groups have been much discussed. The simplest idea—enforced geographic isolation—remains the favorite to this day. This notion boils down to the fortuitous isolation of a portion of a species by an accident of changing geography, or the accidental introduction of an organism (one pregnant female or a single wind-blown seed could be enough) to some remote place not normally inhabited by the species. An example of the former is the emergence of the Isthmus of Panama above sea level, isolating the Caribbean from the Pacific only some 4 million years ago. Colonization of some newly created volcanic oceanic islands by various plants and animals, such as Darwin's finches on the Galapagos and the lemurs of Madagascar, exemplifies the process of isolation by accidental changes in the distributions of organisms. In either case, the net effect is the same: enforced isolation. Thus forcibly fragmented, each subgroup may accrue changes in behavior, anatomy, or genetics. To a certain extent, some divergence will inevitably occur simply on the grounds of probability alone: the two groups would start out with different proportions of the genetic properties of the ancestral species, and random factors of inheritance themselves will accentuate the differences between the two groups. In addition, given such isolation the opportunity for natural selection to modify the adaptations more precisely to the details of the newly occupied local environment—the basic principles of neo-Darwinism we have been discussing—would be likely to occur. So far, so good: the origin of species fits in nicely with the synthesists' reduction of all evolutionary change to natural selection modifying adaptation.

But the problem with this reductionist argument is that reproductive isolation between two groups, formerly united into a single species, cannot be looked upon as adaptively advantageous per se. There is no evidence, nor any reason to suppose, that natural selection acts directly to create reproductive isolation. Full reproductive isolation does not hinge on the degree of adaptive difference between two fledgling species. A species can be adaptively differentiated to an amazing extent, yet its far-flung individuals readily mate on contact and, more important, maintain a network of genetic togetherness through sporadic contact. Conversely, two species can be virtually identical, yet reproductively be completely isolated from each other. Thus mere geographic isolation itself has four possible outcomes: (1) great divergence, plus reproductive isolation; (2) little or no adaptive divergence, but reproductive isolation nonetheless; (3) great adaptive diversity, but no breakage into two or more reproductively isolated groups (new species); and (4) little or no adaptive divergence and no establishment of new reproductively isolated groups. The former two possibilities create new species, the latter two do not. And of the two possibilities for speciation, only one involves significant amounts of adaptive behavioral and anatomical evolutionary change.

In addition to fortuitous, enforced geographic isolation, there are other ideas on how reproductive isolation—the origin of new species—might occur. In these, the adaptive model for the origin of new species fares if anything even worse than in the case of geographic speciation. For instance, chromosomal rearrangements can become established rather suddenly within a population (the classic case being in certain flightless Australian grasshoppers). Such rearrangements can make pairing of chromosomes (one from each parent) simply impossible, creating instant reproductive isolation! This and other recently discussed means of achieving reproductive isolation create the possibility of adaptive change within newly created species, which is precisely the reverse of conventional thought. The idea that full reproductive isolation may in some instances be a prerequisite for adaptive change to occur is an ironic twist to the argument that the origin of species entails something more than the adaptive modification of an ancestor into one or more descendants.

We are now in a position to explain our earlier remark that evolutionary biologists have spent far more time worrying about "modification" than "descent" insofar as Darwin's concept of evolution as "descent with modification" is concerned. Speciation—the origin of new reproductive communities from old—is the process of ancestry and descent. As we have seen, Darwin did not discuss the origin of new species. He couldn't. He had to show that species are ephemeral merely to establish the very notion of evolution. He invented the myth that species were not real to convince the world of the nonmyth that evolution had occurred. Ever since, the notion of evolution solely as *change* has been paramount in the minds of the vast majority of those who have thought about the evolutionary process. And as a result we now have a fairly sophisticated grasp of how behavioral, anatomical, and genetic change occurs in evolution. To a large extent, it *is* adaptive, and natural selection does remain an effective means of accounting for adaptive change in evolution.

But in trying to explain the origin of new species from old—or why there are so many different kinds of creatures in the world today—as a simple by-product of our theories of change, we have virtually ignored the fact that species are integral wholes: communities tied together over space and through time by patterns of reproductive continuity. Descent, the origin of new species from old, is only tangentially related to change, and is certainly not a simple consequence of the processes which give us evolutionary change. Sometimes we see speciation with much change; sometimes we see both little change and no speciation. But we also see speciation with little change, and even a great deal of change with little or no speciation (though, interestingly, the latter pattern is more often encountered in geographic variation than in patterns of change through time).

So, instead of the simple process of adaptive change that is the stuff of the modern synthetic theory of evolution, we have apples and pears: two ingredients which though related are not the same. The significance of all this is that Darwin's prediction that long-term evolutionary change should produce a systematic pattern of gradual, progressive change in the fossil record was faulty. Viewing species as discrete, individual units, each undergoing its own separate evolutionary history (but which, from time to time, give rise to new, descendant species) leads to a rather different prediction of what patterns of change in the fossil record should really look like. And, given the extraordinary stability that is the rule among fossil species, we might expect to find changes in the distribution of organisms— the places where they live—or their outright extinction, rather than any real evidence that animals and plants react to changing environments by sitting still, grinning and bearing it, and adapting to meet the exigencies of ongoing environmental change. How do patterns of evolutionary change actually look in the fossil record?

CHAPTER FOUR

Patterns Great and Small: Evolutionary Change and the Fossil Record

CHANGE IS the basic message of the fossil record. Sediments left by ancient seas, lakes, and streams have hardened into a great pile of rocks to form the outer skin of the earth's crust. Where erosion has cut into these rocks, we see a long sequence of successive layers, like the pages of a book, recording the events of earth history, sometimes in extraordinary detail.

Some of the layers are fairly teeming with fossils. Others are nearly barren. But the fossil record, though far from perfect, presents a compelling pattern of change in the development of life that cannot be ignored. Fossils of complex forms of animal life first appear in rocks about 600 million years old. Mile upon mile of sediments have accumulated over the oldest of fossiliferous sedimentary rocks. Throughout these layers lies a pattern of change: disappearance of earlier forms, appearance of new. Some creatures, such as horseshoe crabs, have survived to the present day in more or less unaltered form since their ancestors first made their appearance on earth—in the case of horseshoe crabs, some 300 million years ago. Other groups are of more recent vintage: mammals first appeared midway through the Mesozoic, the heyday of dinosaurs; horses appeared in the Eocene, the geologic epoch spanning the interval of 55 million to 38 million years ago. The celebrated *Eohippus* (dawn horse), more accurately called *Hyracotherium,* is the oldest horse so far found. The size of an average domestic dog, dawn horses resembled the generalized mammals of that remote time far more than do our modern, advanced, horses. Their teeth were very similar to those of their contemporary close relatives who were to become ancestral to rhinoceroses and tapirs, the extinct titanotheres, and the chalicotheres. Horses, rhinos, titanotheres, tapirs, and chalicotheres are a natural evolutionary group, all descended from a single common ancestor. The genealogical production of this array was recognized far back in human history: tapirs, rhinos, and horses (including wild asses, donkeys, and zebras) have an odd number of toes on their feet—modern horses having but a single toe. Deer, cows, pigs, sheep, and antelope—the mammals most likely to be confused with horses and rhinos as large herbivores—have an even number of toes—two, the cloven-footed animals of the Bible. We call the odd-toed group perissodactyls, the even-toed group artiodactyls. And more similarities than mere toe number unite each of these groups.

One would predict that progressively older fossils should resemble the primitive, ancestral condition more and more closely. Eocene perissodactyls should look more like the common root of the entire stock than do the widely different looking modern horses, tapirs, and rhinos. And they do. Dawn horses had three toes on the front feet, and four on the hind feet (the odd-number scheme had not yet become fully entrenched back then). The primitive mammalian complement is five fingers and toes, the number we ourselves retain.

As we trace fossils from a single lineage, such as the horses, up through the rock sequence, we should be able to see them assume their present form. We do. Horses are an excellent example because they have left a dense fossil record. Anyone untrained in biology or paleontology, if asked to line fossil horses up according to their degree of "modernness" (in size, complexity of dentition, number of toes, and a host of other features), would put them in the proper order of their geologic age knowing nothing of the actual geologic positions of the various fossils. The only source of confusion is the occasional persistence of an ancestor after its descendant has come and gone. This is not to suggest that the reconstruction of evolutionary history is a casual affair: horse evolution was apparently labyrinthine, with many twigs and side branches on the family evolutionary tree. It takes someone skilled in mammalian anatomy to unravel the details and precise course of horse evolution. But as far as the basic aspects of change in horse evolution over the past 50 million years are concerned, the simple truth is that the older the rock, the more primitive the fossil horse we'll find in it.

Thus the overall picture presented by the fossil record confirms the most basic predictions we can make to test the very notion of evolution: if all organisms are related by a process of ancestry and descent, older rocks should contain more primitive members of a group than younger rocks. We should be able to document progressively more advanced forms as we look in correspondingly younger rocks. This is what we find.

But this very confirmation of the most basic of evolutionary predictions has led us astray. As we have seen in the previous chapter, the usual conception casts evolution as a gradual, steady process of adaptive change. And we have already seen that the fossil record conflicts with that view. Now let's look at the fossil record to see what patterns of evolutionary change are actually there. The general agreement that older rocks produce more primitive fossils and that as we look in younger rocks we usually find more advanced members of an evolving lineage has been taken as sufficient evidence that the evolution of life is fundamentally a process of gradual, progressive, adaptive change. But when we take a second, harder look at the fossil record we begin to see the truly mythic qualities of this story. For the gross patterns of evolutionary change so abundantly documented in the fossil record could have been produced in a number of different ways. We are faced more with a great leap of faith—that gradual, progressive, adaptive change underlies the general pattern of evolutionary change we see in the rocks—than any hard evidence. In fact, a closer look at the fossil record shows that another view, centering around the evolution, stability, and death of individual species, predicts a pattern of change that fits the facts of the fossil record much more closely.

The notion of gradual, progressive change collides head-on with the stability seen in most fossil species, for the general progressive sequence of life's evolutionary history seen in the fossil record has always been taken as confirmation of the underlying assumption that all change comes from progressive generation-by-generation modification of species. What the record is really telling us is that evolution, as suspected, has occurred. But we have greatly erred in predicting what the pattern of change should look like in the fossil record. Rather than taking the record literally, we have dismissed the lack of change within species as merely the artifacts of an imperfect record. But the time has come to ask, instead, if the record isn't telling us something that our theories ought to be able to explain—rather than explain away.

Gaps: Real and Artificial

The fossil record is full of holes. No species' record is perfect, because in no species has more than a minute fraction of its constituents become fossilized. Even the very best records—those left by minute marine organisms whose remains litter the sea floor, accumulating by the billions in the bottom muds—are shot full of holes. That there are gaps within the fossil records of species has, in a curious way, comforted Darwin and his successors: the lack of intermediate fossils, "missing links" between species, genera, families, and the higher groups of the Linnaean hierarchy, is seen as the artifact of the imperfect fossil record.

But nature abhors a vacuum. The fossil record of complex life began some 600 million years ago. It is punctuated by many episodes of extinction, some so massive that life was drastically reduced to a fraction of its normal diversity. The greatest of these ecological crises ended the Paleozoic and Mesozoic eras.

But many other events, some only slightly less devastating, have shaken up our planet's ecosystem over the past half billion years. And after each great dying, land and sea were repopulated by a new array of species, basically similar to their predecessors but different in detail. Over the eons, some groups disappeared entirely—whole orders, classes, and even phyla. And in their stead came new groups.

One striking aspect of these extinction/rebound episodes in life's history is the extraordinary rapidity with which they occur. The Cretaceous extinction about 65 million years ago, which took away the last of the dinosaurs and perhaps as much as 90 percent of all the other forms of Cretaceous life, took place within the span of a million years. Now, a million years is certainly a long period of time by some standards, but it is an eyeblink in geologic history. Events occurring within less than a million years' time can create patterns of abrupt change in the fossil record: in many places around the world, fossils can be traced up into the highest layers of Cretaceous rocks when, all of a sudden, they just disappear. And the rocks immediately above preserve representatives of the initial repopulation, life's rebound after the collapse. The record jumps, and all the evidence shows that the record is real: the gaps we see reflect real events in life's history—not the artifact of a poor fossil record.

Speciation can occur very quickly. In perhaps a few hundred years, new reproductively isolated species can form. All that is required is the opportunity to do so. Invasions of new habitats unoccupied by competitors offer such opportunities. When the first "finch" arrived in the Galapagos Islands from the South American mainland, it encountered a habitat rich in opportunities. The variety of ecological niches was soon exploited by speciation; the ancestral finch produced descendant species whose adaptations differed from the ancestor's as the various unexploited niches were seized upon. Such a pattern of speciation, where an ancestral species quickly gives rise to a number of divergent forms, all occupying different niches, is called "adaptive radiation."

This opportunism, this penchant for seizing new opportunities by spinning off new species rapidly (rather than by the gradual transformation of an entire ancestral species into one or more descendants) is perhaps the most basic, fundamental pattern of the history of life. Once complex animal life had come into being those 600 million years ago, it took only about 15 million years to establish all the basic groups, or phyla, of animal life. Rapid diversification in previously unexploited habitat space gave us the basic elements of animal life in one tremendous explosion. Thereafter, exploitation of new environments (such as the invasion of the land by plants, vertebrates, mollusks, and arthropods) or, more commonly, episodes of reproliferation following extinctions produced all the subsequent events of major evolutionary change.

The moral is this: once a species appears and occupies a niche, it will continue to hold on to that piece of ecological space until forced to relinquish it. The habitat itself may change too much, or some other species able to exploit the environment more efficiently may eventually eliminate or outdo a species. But until this happens, species maintain a tenacious grip on their niche. Consequently, ecosystems are extraordinarily conservative. Turnover is slow: there is little opportunity for evolution in a mature ecosystem. There is little or no reason for component species to change, and it is difficult for new species to gain a toehold, the necessary inroad to insure long-term survival. The common pattern is for new species to take advantage of the hard times of others. Extinction causes vacancies, and there are always plenty of applicants to fill them. Mass extinctions cause rapid, wholesale turnovers. But there is also a steady background ticking of extinction and speciation turning over the constituent species of the earth's ecosystem.

Thus it is speciation which allows change to happen. And because successful speciation can be quite rapid, and is sporadic in its frequency, the pattern of change in the fossil record should be episodic. The late P. C. Sylvester-Bradley said it best when he likened evolution to the life of a soldier: long periods of boredom punctuated by brief periods of terror. There is little or no change until the previous equilibrium of the ecosystem is disturbed, soon to be replaced by a new equilibrium achieved by a newly organized set of players. Thus, the evolutionary game is governed by ecological rules, not by the rules of genetics; and the players are species, not gene frequencies within species.

Rapid speciation, by the fragmentation of one reproductively integrated species into two, produces the stasis-followed-by-rapid-change pattern we see exhibited between closely related species in the fossil record. New large groups, such as mammals and birds, appear and become successful as ecological opportunities of sufficiently great magnitude appear. Birds invaded the air, and with feathers serving to maintain a high level of internal body heat as well as a more effective means of flight, were more successful than the pterosaurs, which had achieved flight earlier. Mammals appeared in the Mesozoic and proliferated into several major groups (all but one of which are now extinct); but they became truly established only when the Cretaceous extinction took out the ruling reptiles, opening the door for the tremendous radiation of mammals in the Lower Tertiary.

All patterns great and small in the fossil record thus reflect the basic economy of nature: there is finite room, a finite number of niches, in an ecosystem. Ecosystems, quickly filled, are conservative; they are in rough equilibria, as all change, major and minor, comes from the deletion and addition of species. The gaps between species, and the gaps between phyla, all reflect these few simple truths. Speciation, ultimately dependent on niche availability, allows change to happen. The myth that change itself produces new species is gone. Instead it is new species that produce change.

Evolutionary Change: Random or Directed?

One final word about patterns of change is in order before we ask how well the physical evolution of our own lineage fits the patterns typical of other organisms. The basic Darwinian thesis that apparent design in nature comes about by natural causes, specifically a process of adaptation by natural selection, remains intact. The mythic aspect of the notion of adaptation produced by natural selection comes simply from extrapolating the process too far: insisting that this sort of change alone is what evolution is all about leads to the notion that natural selection is constantly at work, slowly but steadily modifying a species' adaptations to produce all the aspects of life's history. It leads to expectations of pervasive gradual change. The myth is debunked by the pattern we see, which contradicts the smoothly progressive, constant pattern of change we expected. Speciation is the key, but adaptation via selection is not thereby ruled out as the underlying agent of change. Rather, speciation is what triggers all but the most trivial examples of adaptive change.

Thus adaptation and selection remain at the core of our explanation of the actual anatomical and behavioral changes we see in the fossil record. And this is a typically deterministic stance: the change we see in evolution is not random; it comes about primarily as a response to real-world conditions. Repeat the "experiment" under the same conditions; start the evolutionary clock all over again, say with *Hyracotherium*, the earliest known horse, and substantially the same sorts of horses will appear. Of course, the further back we reset the clock, the more different the end results will be. And this factor introduces the chance element into the evolutionary game.

Speciation effectively disrupts the pattern of parental ancestry and descent that binds a species together. Since natural selection is differential reproductive success from generation to generation, speciation disrupts patterns of selection within species, and this agrees well with the observation that speciation is not a function of adaptation, but that large-scale adaptive change relies, instead, on speciation. So evolution proceeds at two different levels: within species and between species. Within species change, though not as ubiquitous, inevitable, or as consequential as once imagined, occurs deterministically by natural selection. But genetic drift, a sampling effect which amounts to luck, also enters in. Krill (oceanic shrimp) strained from the sea by a huge cruising baleen whale die, while their luckier cohorts outside the whale's path survive. The death of the krill unfortunate enough to be in the whale's path has nothing to do with inferior adaptations: it is simply bad luck.

Precisely the same sort of random factor can exterminate whole populations, and even entire species. Species of land snails on Pacific islands are frequently restricted to single valleys. Closely related species may occur sequentially in the erosional valleys ringing a volcano forming such an island. An eruption, sending a stream of lava down one side of a volcano, could extinguish one species of snail but not its close relatives in adjacent but lava-free valleys. Extinction here would be deterministic in the limited sense that the exact cause of death was that all individuals of the species were covered by a blanket of molten lava. But why one species survived and another did not had nothing to do with inherent superiority of the surviving species over its newly extinct relative. It was just luck, irrelevant to any differences in adaptation between the two species. Thus although the extinction of a species has a definite cause, we may find an element of randomness when we come to decide how it is that some species survive while others of the same lineage do not.

But we must also ask if there is a nonrandom element here, operating between species just as natural selection works within species. Do some species in a lineage survive while others become extinct because of a nonrandom selective process? The answer appears to be yes: species selection, the differential survival of species based on properties of the species themselves, seems to be a reality. In fact, species selection appears to be the mechanism underlying most of the large-scale evolutionary patterns we see in nature.

We have already mentioned adaptive radiations, the rapid proliferation of new species every which way as a response to the invasion of virgin ecological territory with a host of new opportunities. Such radiations are perhaps the most important of all large-scale evolutionary patterns. But equally interesting and in some respects more compelling evidence for the existence of species selection is so-called trends: the tendency in many lineages to show anatomical change in a unidirectional fashion over the long course of the group's evolutionary history. Trends in horse evolution, for example, include reduction of toe number, increase in body size, and the progressive increase in height of the molar teeth and complexity of their ridge patterns (for grinding grass) among the grazing horses. Trends, of course, have always been cited as the best fossil evidence for Darwinian gradualism. But critical looks at all the best-known examples invariably show that the trends are *between* species. Within species there normally is no progressive modification by natural selection. Trends thus appear to reflect nonrandom interspecific survival and "reproductive success," since it is the surviving species which leave new, descendant offspring in the future.

Nor are trends the simple, linear arrays of ancestors and descendants they once appeared to be. Rather, the anatomical change that accompanies speciation may very well go in a direction opposite to that of the prevailing trend characterizing the lineage as a whole. Suppose, for instance, a trend of size increase characterized a lineage throughout its history—as happened, for example, among horses in their long evolutionary history, and indeed in our own lineage. Fossil species of horses and humans show no evidence of progressive size increase during their own histories. Nor must every descendant species be larger than its ancestor. For instance, it is only as a long-term net effect that, on average, larger species outlived and outbred smaller species. What superficially looks like a smooth, linear increase in size was actually an episodic culling process, apparently the result of selection acting on entire species. Evidently not random, the trend also cannot be explained by pure natural selection, as no progressive adaptive change in the direction of the trend is ever seen *within* species—the level at which strict natural selection operates.

This alternative view of evolution, which sees an interplay of random and deterministic processes occurring at two different levels that are mostly if not completely divorced from one another, undoubtedly has mythic aspects of its own. At the moment the reductionist view that all evolution is a simple function of adaptive change via natural selection has been exposed as inadequate. This older view is distinctly Victorian, built as it is upon the larger myth that change is inevitable, progressive, gradual, and always for the good. The newer view, while retaining the ingredients of selection and adaptation, is a larger, more embracing scheme. It admits descent more fully into the equation: evolution = descent + modification. And it better fits the facts, the patterns we see in nature.

But this newer view must itself inevitably give way someday to an even more accurate estimate of the true nature of the evolutionary process. Ironically, we urge here a view of the inevitability of change. Ideas, or sets of ideas ("paradigms"), remain in favor for a period, during which a critical mass of anomalies builds up. Suddenly the old paradigm is discarded, and the new one takes its place. To date, no set of ideas about anything has remained inviolate throughout human history. Change of ideas so far seems inevitable. This includes our most fondly cherished ideas, even though individuals themselves rarely change their minds. It is their intellectual descendants who do the changing. But Thomas Kuhn's idea of paradigm shifts in the history of science, the view we just briefly characterized, does give us a curious analogy with the view of evolution we have presented: both see change as coming in fits and starts. This is not the only time we will encounter similarities in patterns of change in two otherwise utterly different systems. But we get ahead of ourselves. We must now look at the fossil record of human physical evolution and then ask which view of the evolutionary process best explains the historical pattern we find.

CHAPTER FIVE

Fossils and Finders: The Cast of Characters in Human Evolution

BEFORE WE embark on our attempt to discover whether or not the events of human evolution fit the pattern we have observed elsewhere in the evolution of life on earth, we should summarize briefly the highlights of the history of the hominid fossil record to introduce, if not the entire cast of characters, at least the stars.

The Neanderthals

The science of paleontology had had a turbulent if not uniformly glorious history before Darwin's publication of the *Origin of Species;* but those few documents which had already been discovered of the human evolutionary past had remained obscure or unappreciated. Actually, the first such evidence to come to light came not in the form of fossils but of stone tools, when in 1797 the English antiquarian John Frere recognized that the shaped flints often found in southern English gravel pits were the work of men who lived at a time "even beyond that of the present world." Frere left it at that, however, and it was not until thirty years or so had passed that a French customs official named Jacques Boucher de Perthes began pointing out that flint implements found in the gravels of the

Somme River, in association with the remains of extinct animals, constituted incontrovertible evidence of the former presence of ancient man in northern France. But, "l'homme fossile n'existe pas" had said Georges Cuvier, and Boucher de Perthes was almost universally ignored or derided. Cuvier was the foremost proponent of the idea that a series of catastrophes, the most recent of them being the biblical Flood, was responsible for the succession of extinct life found in the fossil record; and although he died in 1832, his hand for long lay heavy over French science.

The fossils, such as they were, fared little better. In 1848 a skull was recovered during quarrying operations on the Rock of Gibraltar. It found its way into an obscure local museum, where it languished for fifteen years before being dispatched to England. A skeleton found nine years later in the Neander valley, near Düsseldorf, Germany, received more attention, but initially at least, most of it was unfavorable. The original Neanderthal, this specimen consisted of a skullcap, lacking the face and skull base, together with some limb bones. Since it was unaccompanied by any direct evidence as to its antiquity, the specimen could be evaluated only on the basis of its morphology—its physical appearance. And this, at least as far as the skullcap was concerned, was highly unusual. In contrast to the high-vaulted, smooth-browed, lightly built skull of modern man, the Neanderthal cranium was long and low, protuberant at the back, and adorned in front with heavy browridges. After an initial hue and cry when its describers interpreted the skeleton as that of an ancient and barbarous European, comment more or less lapsed until 1861, when its description was translated into English, with additional comments by the translator, George Busk. Busk, a proficient anatomist, pointed to resemblances between the Neanderthal skull and that of the then newly discovered gorilla, and in doing so unleashed a torrent of debate.

Essentially, the argument resolved itself into two camps. On one side were those who considered the peculiarities of the Neanderthal skull to represent no more than pathological alterations to a modern specimen. The most elaborate version of this theory was put forward by the anatomist Friedrich Mayer, who suggested that the unfortunate individual had suffered since childhood with rickets and a broken elbow. The constant frown brought on by the pain of these disabilities, Mayer said, had caused the formation of the browridges. Noting that the limb bones, although robust, resembled those of a modern bowlegged man, and believing that the skull showed Mongolian affinities, Mayer neatly put these observations together in the suggestion that the skeleton was that of a sick Cossack, belonging to the Russian invading force of 1814, who had deserted and had crawled into the cave to die! A less imaginative but more common theory was that the skull had belonged to some kind of pathological idiot, and affinities were sought everywhere. A Dr. Prunier-Bey, for instance, declared the skeleton to be that of a strongly built Celt, "somewhat resembling the skull of a modern Irishman with low mental organization."

Arrayed on the other side of the argument, and by now with the work of Darwin to lean on, were those who believed that Neanderthal man did indeed represent a link with an earlier time in human evolution, combining features of modern man with attributes of his closest living relatives, the great apes. Foremost among these, almost inevitably, was the redoubtable Thomas Henry Huxley, "Darwin's bulldog," the man who had taken up the cudgels for evolution when Darwin's retiring disposition and intellectual caution had removed him from the more public areas of the fray. Huxley emphasized the size of the brain which had been contained within the Neanderthal skullcap; at about 1,200 cc., it fell well within the modern human range and vastly above that of even the largest ape. This raised the tricky question of the intellectual abilities of Neanderthal man. He wasn't a modern idiot, but could a creature with such apelike features possibly have had the sensibilities of modern man? The smart money said no, and even William King, who admitted the specimen into our own genus as *Homo neanderthalensis*, described him as "benighted."

At this point, clinching evidence that the Neanderthal fossil was no aberration turned up in the form of the arrival in England of the Gibraltar skull. Its description fell to Busk, who delighted in speculating that "even Professor Mayer will hardly suppose that a ricketty Cossack engaged in the campaign of 1814 had crept into a sealed fissue in the Rock of Gibraltar." Busk emphasized the similarities between the Gibraltar and Neanderthal specimens, and must have been greatly disappointed when in 1872 the immensely influential German biologist Rudolf Virchow returned to the notion of pathology to explain the anatomical peculiarities of the Neanderthal skullcap. It was only after many more skeletons showing essentially the same characteristics had been discovered (for example, at Spy, Belgium, in 1886; at La Chapelle-aux-Saints, France, in 1908; at La Ferrassie and La Quina, France, in 1909 and 1911) that the former existence of a distinct Neanderthal population became incontrovertible. Meanwhile, Virchow's influence, together with the discovery of genuinely modern-looking fossil humans, was distracting attention from the Neanderthals as the putative makers

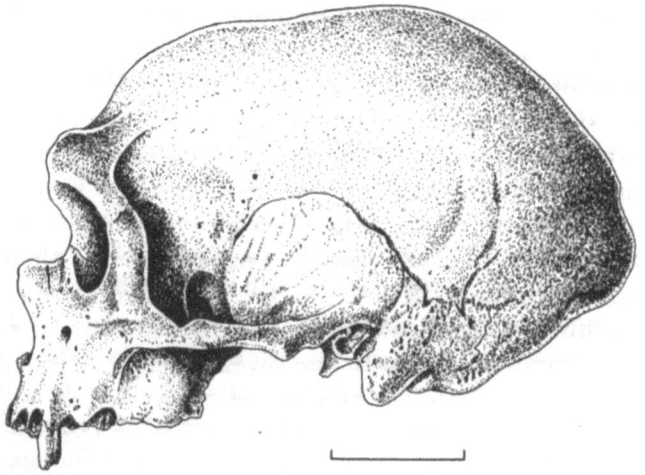

A Neanderthal cranium: the "old man" of La Chapelle. The scale represents five centimeters.

of the stone tools which archaeologists, acting at last on the initiative of Boucher de Perthes, were beginning to recover in abundance.

The most famous of the modern-looking fossil humans were found in the rock shelter of Cro-Magnon, a locality near Les Eyzies, southwestern France, in 1868. There was at the time some dispute over whether the five Cro-Magnon skeletons represented burials or not, but nobody doubted the association of the fossils with the remains of extinct animals and of Old Stone Age tools, and their antiquity was unquestioned. In this scheme of things the Neanderthals were anomalous, and as discoveries of Cro-Magnon type continued to accumulate, the Neanderthals were increasingly dismissed as oddities. As it happened, the Cro-Magnon finds antedated any finds of Neanderthal types in actual association with extinct animals, and this historical accident served to reinforce this attitude.

At this juncture, it is only fair to pause to point out that everyone, pro-Neanderthal and anti, was working in what was largely both an intellectual and a factual vacuum. Not only was Darwinism far from being universally accepted, particularly in continental Europe, where the fossils were found, but there existed as yet absolutely no paleontological framework within which to place these fossils. What seem to us now to have been ludicrous flights of fancy did not necessarily have that flavor a hundred years ago. The job of science, then as now, was to explain facts. And the new facts represented by the Neanderthal and Gibraltar discoveries bore no relation whatsoever to any existing bodies of knowledge; there was simply no place for them to be fitted into. Today, we have almost the opposite problem in interpreting new hominid fossils; often, there are too many places into which they might potentially fit. But we, at least, are not groping totally in the dark. Newton claimed to have seen further only by standing on the shoulders of giants; and the perspective we can now bring to our attempts to understand the evolution of mankind has been achieved only by standing on the rubble of literally hundreds of collapsed theories, and by the remarkable augmentation of the fossil record—the paleontological framework—that has taken place this century.

Java Man

One of the most incredible good-luck stories in the entire annals of paleontology—and good luck plays no small role in hominid fossil-finding—must be that of Eugene Dubois. In his *Descent of Man,* published in 1871, Darwin had suggested that man had originated in the tropics, most probably in Africa, the continent harboring two out of his three closest surviving relatives, the great apes. Dubois, a young physician who had collected fossils in his native Holland since his childhood, became obsessed with the idea of finding fossils of early man. Unable to raise funds privately to support an expedition to the tropics, as a last resort he abandoned his career as an anatomist and signed up as a medical officer in the Dutch East Indian Army. In 1887 he sailed with his family for Sumatra. The East Indies were not Africa, but they were in the tropics, they did belong to Holland, and they were the home of the orangutan, the remaining member of the great-ape triumvirate. Moreover, the eminent German embryologist Ernst Haeckel had recently emphasized the resemblance of certain characters of the gibbon, a lesser ape, to those of man. And the gibbon is found in the East Indies.

One of the aspects of Haeckel's work which most appealed to Dubois was his reconstruction of a "chain of animal ancestors of man." Haeckel had been greatly impressed by the similarities of the apes to man, but to his mind one distinction was paramount: humans have articulate speech, apes do not. The apes, Haeckel claimed, were the penultimate living link in the chain leading from the simplest forms of life to ourselves. But the link between the apes and true man—"ape-like man," bereft of the powers of speech—was missing. To this hypothetical link Haeckel gave the name *Pithecanthropus alalus* ("ape-man without speech"); and it was this link, the halfway form between modern humans and modern apes, that Dubois was determined to find.

Dubois quickly enlisted the support of his superiors in his quest for fossils. After two years of fruitless search in Sumatra he moved to Java, armed with official instructions to carry out paleontological investigations at his own discretion. Almost all human fossils yet recovered had been found in caves, and at first Dubois gave his exclusive attention to exploring caves in southeast Java. But nothing he found was old enough to be of interest to him, and he soon turned to the local sedimentary sequence. Sedimentary rocks are composed of particles derived from other rocks, and they are the only rocks which yield fossils. Layers of sedimentary rocks are formed when such particles are transported from their place of origin by wind or (more commonly) water, and are laid down elsewhere. This happens when the stream or whatever is transporting the sediments runs out of energy, for instance when it flows into a lake and its movement is dissipated. Over a period of time more and more particles will settle out of the water and a layer of sediment results. If the sediments become buried deeply enough, the combined effects of compaction and cementation by filtering minerals may harden them to the consistency of rock. Many layers may accumulate in this way, their characteristics depending on the source of the particles comprising them and on the circumstances of deposition.

Under the right conditions fossils may be formed when the bones of dead animals are covered by sediments before they have been destroyed by carnivores or the elements. Eventually the bones may become "permineralized": hardened by the addition of minerals. Obviously, in the absense of major disturbance, layers higher in the sequence will be younger in geological age than those lower down, and so will any fossils contained in them. Should erosion then cut down through the sediments and re-expose the fossils at the surface, those which erode out higher in the section will thus represent life forms which existed later in time than those which erode out further down.

Very soon Dubois' excavations, carried out with the help of convict labor, began to produce copious quantities of fossils—of fish, reptiles, and mammals of many kinds. These fossils suggested that the strata were of middle Pleistocene age, thus older than any in Europe which had yet yielded fossil hominids. But only one hominid specimen, a rather indeterminate fragment of lower jaw, turned up, and after a while Dubois moved the focus of his activities to another section of the deposits, some fifty feet thick, which had been exposed by the downcutting of the Solo River. Here his workers hit paydirt in short order. In September 1891 a human tooth turned up, and in the following month a skullcap. At first Dubois thought he had the remains of a chimpanzee, but the skullcap was unlike anything ever known before. Its bone was thick and heavy, its braincase long, low, and poorly filled out, with a capacity of only 900 cc. (the average for modern humans is about 1400 cc., versus about 400 cc. for a chimpanzee). At the back the cranium was sharply angled in side view, and at the front it bore distinct, inflated brow ridges. The next year another tooth and a femur (thighbone) were added to the collection. The femur looked to all intents and purposes like a robust modern bone, its only remarkable feature being a large pathological outgrowth. It was found at the same stratigraphic level (i.e., in the same layer of the deposits) as the skullcap but some distance away from it.

As he studied these prizes, Dubois found his interpretation changing. He rapidly formed the belief that the bones were those of an extinct manlike ape, transitional from the great apes to man. Partly in homage to Haeckel, and partly to reflect the erect striding locomotion so clearly indicated by the femur, he called this new form *Pithecanthropus erectus*.

In 1895, by which time general scientific opinion was becoming more receptive than it had been to the idea of "primitive" proto-humans, Dubois completed his military service and returned to Holland bearing his fossils. Eagerly he presented them, and his conclusions about them, to the scientific world. The occasion was a meeting chaired by Virchow, and as might have been expected, the august gentleman refused to accept that the skullcap and the femur were properly associated together. Worse, even the admired Haeckel was heard to remark on the inadequacy of the specimens. Few agreed with Dubois, and the common reaction, at meeting after meeting, was that the femur and the skullcap belonged neither to the same individual nor to the same species. Only in England did those who counted, notably the anatomist Arthur Keith, come to see truly human affinities in the material.

Rejection spurred Dubois to strive ever harder to prove that both specimens were intermediate between ape and man: not such a difficult feat in the case of the skullcap, but greatly so in the case of the femur. As the argument continued, Dubois' frustration at failing to find converts increased until eventually he withdrew from the scene, taking his fossils with him. For a quarter of a century he remained almost silent on the subject, the fossils buried beneath the floor of his dining room. When he finally brought them to light once more, in 1923, to show them to some visiting American scholars, his ideas had radically changed. While his visitors concluded that the bones really were those of an early form of man, Dubois responded that they had in fact belonged to a giant gibbon.

Despite the rapidly accumulating evidence of their human affinities, Dubois maintained this absurd belief right up to his death in 1940. Today Dubois' Java man, known now from several more specimens, including a skull with a face, is unanimously admitted to our own genus as *Homo erectus;* some other specimens from the area may represent an older form yet.

A Fallacy and a Fraud

While the debate over Dubois' Java man was still echoing, fossil material continued to accumulate from sites in Europe. Perhaps the most significant of these, certainly in terms of its public and scientific impact at the time, was the Old Man of La Chapelle-aux-Saints. Discovered in 1908, in a cave deposit also containing stone implements of "Mousterian" type, this was a virtually complete skeleton of an aged individual whose characteristics matched those of the original Gibraltar and Neanderthal finds. These remains were turned over for study to Marcellin Boule, the premier French paleoanthropologist of his day, and it is to Boule more than to anyone else that we owe the shambling, brutish image of the Neanderthals immortalized in a thousand comic strips.

In a minutely detailed study occupying three large volumes and published between 1911 and 1913, Boule put forward an image of a monster unable to stand completely erect, who walked on the sides of his grasping feet like an orangutan. His back curved forward, and his head jutted out yet farther. His coarse features reflected an equally coarse and brutish nature. Ironically, whereas Virchow and Mayer had earlier attempted to explain Neanderthal morphology as a pathological condition, Boule reached his conclusions while totally ignoring the abundant evidence that the La Chapelle skeleton did indeed show some pathological alteration. Every feature that Boule stressed in his analysis can be shown to have no basis in fact or to be the result of an advanced case of osteoarthritis. Robust the Neanderthals certainly were; but otherwise their bodies were identical to our own. Distinctions are indeed to be found in the skull, yet even this was balanced atop the spine exactly as ours is. But Boule's authority was close to absolute, and his conclusions strongly affected paleoanthropological thinking for several decades.

Not everyone was convinced, however. At about the same time the German anatomist Gustav Schwalbe proposed two possible pathways of human evolution, based on the relatively few fossils then known and on what was coming to light about their ages. There was, of course, no way to assign an age in years to any fossil, but as the geology of Europe became better known, it became possible to place them in some sort of age sequence. In particular, this was the time during which the geologists Albrecht Penck and Eduard Bruckner were working out the sequence of glaciations in Europe during the Pleistocene, the geological epoch immediately preceding our own

Holocene, which started about 10,000 years ago. Penck and Bruckner defined four major episodes of Pleistocene glaciation, when the polar ice cap expanded to cover much of northern Europe. They called these, from oldest to youngest: Günz, Mindel, Riss, and Würm. A fifth cold period, this one not characterized in Europe by ice cover, has since been added to the beginning of the series.

Between these glacials were warmer periods, called interglacials, when the ice once more retreated, and within each glacial there is evidence for various lesser climatic fluctuations characterized by smaller-scale ice advances (stadials) and retreats (interstadials). Glacial geology is particularly tricky because each ice advance obliterates or churns up the evidence left behind by the one before it, but it was quickly recognized that both the Neanderthals and the Cro-Magnons were late in this scheme of things, and that of the two the Neanderthals seemed to be a little earlier. Boule, of course, deduced that not enough time existed between the coarse Neanderthals and the refined Cro-Magnons for the former to give rise to the latter; clearly, extinction was the only fitting fate for the brutes he had described. Schwalbe, on the other hand, was prepared to entertain all possibilities. His first alternative was to derive both Neanderthals and moderns from the earlier Java man but to represent the Neanderthals as a terminal offshoot which branched off the human line at some time after Java man. His second was to place the three in a single sequence: Java → Neanderthal → modern. It was this alternative that he preferred.

In 1908, a couple of years after Schwalbe's review, Otto Schoetensack announced the discovery of Heidelberg man in a sand pit at Mauer, near the German university town. Although the pit had for years been producing fossil mammals of rather archaic—middle Pleistocene—aspect, the only hominid fossil ever found there—to this day—is the lower jaw Schoetensack described. The particular interest of this specimen, apart from its robustness and its missing chin, was its age: from its associated fauna Schoetensack placed it in the Günz-Mindel interglacial: much older than the Neanderthals.

Schoetensack's discovery won him great and well-deserved acclaim in continental Europe, but little of the importance of his specimen penetrated to England where, in the absence of domestic fossil humans, the leading anatomical lights of the day were expounding theoretically on the course of human evolution, confidently predicting that fully modern-looking human fossils would be found as far back as the Pliocene, the epoch antedating the Pleistocene. One English specimen there was, found at Galley Hill, near London, in 1888. It was discovered in deposits claimed to be of early Pleistocene age, but, since it differed in no way at all from fully modern man, it had been regarded as a relatively recent burial, intrusive into the older deposits. Until, that is, it was rediscovered by Arthur Keith, one of the two most influential British anatomists of his day. Keith affirmed the skeleton's antiquity, and thus its status as evidence of the extreme antiquity of anatomically modern man. To Keith, and to his rival Grafton Elliot Smith, the Neanderthal and Java fossils were no more than aberrant offshoots of the human line, lacking either the large brain or the antiquity to be expected of our ancestor. To Smith, in particular, the possession of a large brain was the outstanding hallmark of mankind: the feature that had both characterized and guided human evolution.

The stage was thus set for the announcement in 1912, by Arthur Smith Woodward, that a large-brained hominid had been found in Pliocene deposits at Piltdown, in Sussex. Woodward was the foremost British paleontologist of the time, an expert on fossil fish and Keeper of Geology at the British Museum. He had received the first Piltdown fragments from Charles Dawson, an amateur fossil collector who had contributed to the museum's collections in the past, and who had himself obtained the specimens from workmen who had been digging in a gravel pit near Piltdown common. Dawson and Smith Woodward subsequently visited the site on several occasions and recovered more material themselves, including mammal fossils indicative of considerable antiquity. Eventually, by the end of 1913, most of the left side and back of a cranium had been recovered, together with the right rear half of a lower jaw which lacked the point of articulation with the cranium, and a single canine tooth. Smith Woodward assembled these various bits into a reconstruction of the entire skull, in which the cranial vault resembled that of a modern man, and the lower jaw was apelike. Proof positive at last not only that man's large brain had characterized his line from the earliest times, but also that the earliest man was an Englishman. The apelike aspect of the lower jaw of *Eoanthropus dawsoni* ("Dawson's dawn man"), as Smith Woodward had named the form, bothered Keith a good deal, but argument (and of that there was plenty) centered on the accuracy of Smith Woodward's reconstruction rather than on the validity of the fossils themselves. And as time went on, continuing finds by Dawson up to his death in 1915 seemed to set the seal on Piltdown's status.

After much debate, the great trio of British paleoanthropology found themselves in substantial agreement. Skepticism remained elsewhere, however, expressing itself primarily in doubts that the Piltdown cranium and lower jaw could have belonged to the same individual. But the antiquity of the large-brained cranium went generally unquestioned, and the combined influence of Smith Woodward, Keith, and Elliot Smith, together with the conversion of many early skeptics by a new reconstruction (Smith's) which made more concessions to the apelike form of the jaw, ensured that Piltdown became the standard by which other hominid fossils were to be measured.

This was unfortunate. As the years passed, fossil evidence accumulating from various parts of the world began to make the Piltdown specimen appear increasingly anomalous. Questions were raised about the dating of Piltdown, and the skull became less and less a standard of comparison. It was not, however, until after the Second World War that Kenneth Oakley applied a fluorine test to the Piltdown material. Buried bones take up fluorine from the surrounding deposits, and do so at a rate proportional both to the concentration of fluorine in the deposit and to the length of time they are buried. Since fluorine concentrations vary from place to place and over time this propensity cannot be used as a means of absolute dating, but measurements of fluorine in fossils recovered from the same deposits can show whether or not they have all been buried there for the same amount of time. Oakley had previously demonstrated that the Galley Hill skeleton was much younger than the mammal fossils found in the same deposits; now he proved the same thing for *Eoanthropus*. This left a significant but as yet unstated problem: the combination of a manlike cranium with an apelike jaw was hard enough to swallow in the remote past; in no way could they be combined in a very recent hominid such as Oakley had shown Piltdown to be. Apes had not roamed Europe for many millions of years, so how had these unnatural bedfellows found their way into the same deposits?

To Joseph Weiner, an anatomist from Oxford, there could be only one answer: deliberate fraud. Together with others, he reexamined the Piltdown specimens. Without doubt, the "fossil" was a plant. The jaw, they found, was definitely that of a recent ape, broken to eliminate those parts which would immediately have given away its identity. The teeth had been filed down for the same reason. The cranial vault was that of a modern human, albeit several centuries dead. The specimens, the (imported) associated fossils included, had all been stained to match the characteristics of the deposits. At last, in 1953, over forty years after its appearance, the chimera of Piltdown was laid to rest. But historically, Piltdown remains important, largely because for so long it impeded the acceptance of genuinely ancient fossil hominids which told a very different story. One of these was *Australopithecus*.

The Southern Ape of Africa

In 1922 a young anatomist named Raymond Dart found himself occupying the chair of anatomy at the Witwatersrand University Medical School, in Johannesburg. Dart had worked for some years in Elliot Smith's laboratory in London, cultivating an interest in the evolution of the brain. During that time Smith was, of course, deeply involved in the brouhaha over Piltdown, and Dart arrived in South Africa with a thorough grounding in the issues at stake in the debate over human evolution. Not that that was what had brought him to South Africa; he simply needed a job. But when, one day in 1924, a student brought him the fossil skull of a baboon she had found in the home of a friend, his interest was immediately aroused. The specimen, he learned, came from a lime quarry at Taung, in Bechuanaland. Fossils such as the baboon skull occasionally turned up in the course of blasting operations. Dart asked a colleague from the Geology department at the university to be on the lookout for fossils while in the region of Taung. The colleague did so and arranged for the quarry management to keep alert for fossils.

Two boxes of rock and fossils duly arrived in Johannesburg. Inside one of them, Dart found something astonishing. What first caught Dart's eye, appropriately enough for a neuroanatomist, was a perfect natural cast of the inside of a tiny braincase. Such casts are formed when the skull of a dead creature becomes filled with sand, dust, and other particles which then harden up as the skull itself fossilizes. The inside of the braincase, which reflects the contours of the brain, is then perfectly reproduced in stone. In this instance, the braincase had been little more than half filled.

But to Dart's practiced eye, that was enough. The specimen was small; it turned out later to belong to an infant at about the same stage of development as a modern six-year-old child. But its brain was big for an ape and, more important in Dart's view, showed external features characteristic not of apes but of man. All that notwithstanding, however, its size was much smaller than current wisdom allowed in an ancestor of man.

Dart dug deeper in the box and found, enclosed in rock-hard matrix, the face and lower jaw into which the brain cast fitted at the front. After more than ten weeks' work of a kind totally unfamiliar to him, Dart managed to expose the face and jaw. What he found, he felt, vindicated his first impressions based

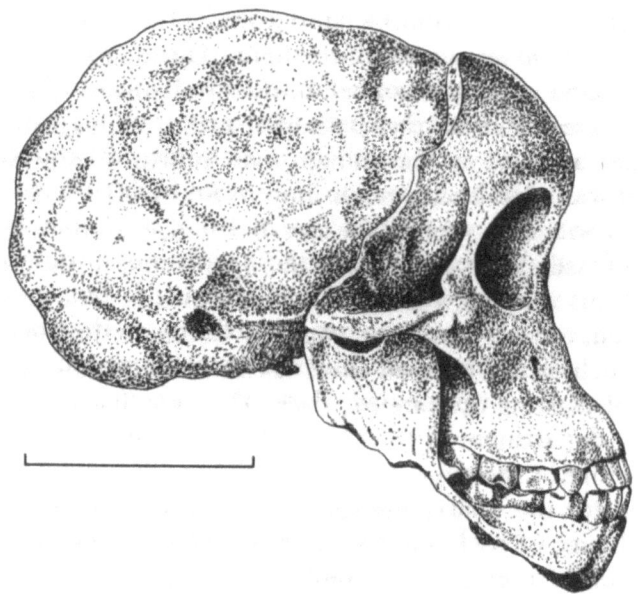

The face and endocranial cast of the Taung child. The scale represents five centimeters.

on the brain. The skull was rather globular, but this is a feature of infants of almost all primate species; the browridges of apes, for example, develop later. But the foramen magnum, the hole through which the spinal cord passes from the brain, was underneath the skull, and not at the back. This central position suggested that in life the creature's head was balanced atop a vertical spine, not hung from the front of an essentially horizontal one as in the apes. Further, the teeth, though large, seemed to Dart to be manlike rather than apelike, although of the permanent set only the first molars were in place.

Dart, alone, with no library, no comparative material, and nobody to discuss his specimen with, ran through the alternatives in his mind. He became convinced that he had before him a "missing link" more central in the story of human evolution than Java man. Dart had, of course, no clear idea of the age of his specimen; all he knew was that its association with extinct baboon species argued for a considerable antiquity. In a paper submitted to the august British journal *Nature,* Dart named his infant fossil *Australopithecus africanus* ("southern ape of Africa"), and placed it in its own zoological family midway between those of apes and man.

His announcement, in early 1925, made an enormous splash in the British popular press but was greeted with reserve by those whose opinion counted. Worse, reserve rapidly hardened into outright opposition. So disheartened was he by this reception that Dart made no attempt to follow up on his discovery by looking for other specimens. He continued work on a monograph describing the specimen, but when in 1931 almost all of it was rejected by the Royal Society, he withdrew from the fray entirely. Why beat his head against a stone wall when the entire paleoanthropological establishment was convinced that his specimen was merely an ape? Besides, the attention of these eminent gentlemen had by now been attracted elsewhere—to China.

Peking Man

After the end of the First World War paleontological collecting in China was more or less monopolized by a Swedish mining engineer and amateur paleontologist named J. G. Andersson, who supplied fossils in considerable quantity to the Paleontological Institute at the University of Uppsala. In 1921, fearing a certain lack of professionalism in Andersson's excavations, the institute sent out a young Austrian, Otto Zdansky, to take charge. Soon after his arrival Zdansky began operations at a fossil-rich cave-fill site near the railroad station at Choukoutien, about thirty miles outside Peking. Before the year was out Zdansky had recovered a fossil hominid tooth at the site, but for reasons of his own mentioned it to nobody, not even to Andersson, who was convinced that quartz pebbles found in the deposits were ancient stone tools, and who had urged him to excavate the site exhaustively. In 1923, Zdansky returned to Sweden to write up his finds, and it was not until

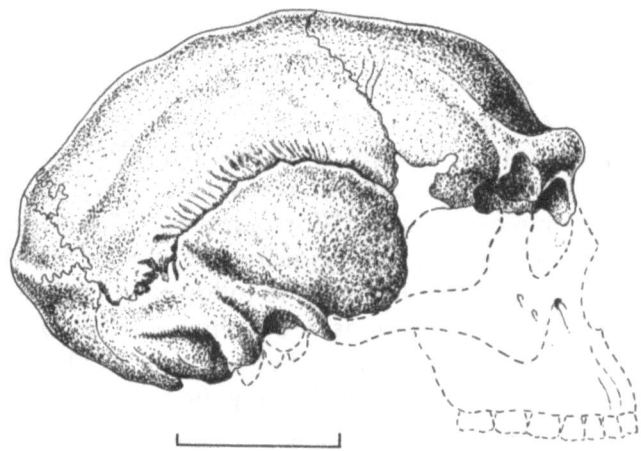

One of the crania of Choukoutien "Peking Man." The scale represents five centimeters.

1926, when Andersson asked for news of spectacular discoveries which could be announced at a meeting to be held in honor of a visit by the Swedish Crown Prince, did Zdansky describe the tooth, plus another he had found while combing through his collection.

At the meeting was Davidson Black, professor of anatomy at the Peiping Union Medical College. Fired with enthusiasm by this news, Black sent a note to *Nature* declaring that this ancient tooth could be assigned to nothing but a species of *Homo*. He next organized a two-year excavation at Choukoutien, funded by the Rockefeller Foundation. Zdansky preferred not to return to direct the excavations; instead, a young paleontologist named Birger Bohlin was dispatched from Sweden to do the job.

The next year's excavation yielded but a single tooth, but in conjunction with the two already known Black felt that this was evidence enough to name a new genus and species of hominid, *Sinanthropus pekinensis* ("Chinese man of Peking"). Most authorities regarded this announcement as premature, and little more than polite interest was generated.

In the following year, 1928, however, something more substantial turned up in the form of a half lower jaw containing three teeth. A tooth and part of a jaw were not a great deal to show for two years of expensive excavations, in which hundreds of tons of infill had been removed from the site, despite the fact that a large collection of mammalian fossils had been made. Black's attention was beginning to wander elsewhere in the quest for hominid fossils, but plans for an alternative fossil-hunting expedition fell through, and the next year, with the Rockefeller Foundation's continuing support, excavation at Choukoutien resumed.

Persistence paid off, and at the very end of 1929 the first skull of Peking man came to light, discovered by the Chinese paleontologist W. C. Pei in a partially filled cave leading off the main excavation. The skull of *Sinanthropus*, represented by a cranial vault lacking the face and most of the base, turned out to be disappointingly reminiscent of Dubois' *Pithecanthropus* from Java, except that the vault was a little better filled out. Subsequent discoveries, continuing beyond Black's death in 1934, together with the meticulous anatomical studies of Black's successor, Franz Weidenreich, amply confirmed this.

By 1940, the cave at Choukoutien had yielded fourteen crania, eleven lower jaws, and an assortment of broken limb bones: the most extensive series of individual fossil hominid crania ever found at a single site. Moreover, the cave had yielded not only stone implements and an extensive associated fauna but also the earliest evidence of the use of fire by ancient hominids.

But America was close to war with Japan, the Japanese were already on Chinese soil, and the political and military situation in China was already confused. In early 1941 Weidenreich left China with a complete set of casts, notes, photographs, and drawings of the Choukoutien hominids. He had been asked to take the original fossils with him but had refused, fearing for their safety en route. There were, equally, fears for their safety if they remained, and at a late stage it was decided to dispatch the fossils to America in the care of a platoon of Marines. With exquisitely unfortunate timing, the Japanese attacked Pearl

Harbor while the Marines were on their way to the port of Tientsin. The Marines were taken prisoner, and the fossils have not been seen since. However, recent excavations at Choukoutien have produced a few new pieces, and work elsewhere in China has led to the recovery of other material of *Homo erectus,* the species to which Peking man is now assigned.

South Africa Again

The Taung skull had been that of a child. Infant apes and humans resemble each other a great deal more than do the respective adults. Until an adult could be found, it was clear, *Australopithecus* was unlikely to get another hearing from the scientific community. That such adults were indeed found was due entirely to the inspiration of one man: Robert Broom, a Scots physician and paleontologist and one of the most colorful characters in the annals of paleontology. Always an academic outsider, supporting himself for the most part by the practice of medicine, Broom had nonetheless become by 1920 the outstanding authority on the mammal-like reptiles found in such abundance in the sediments of South Africa's Karroo. Despite this, it is a measure of Broom's isolation from the South African academic mainstream that in 1933, a decade after he had been made a fellow of Britain's Royal Society and shortly after receiving its Royal Medal in recognition of his outstanding paleontological researches, he was unable to afford the train fare to Johannesburg to address the South African Association for the Advancement of Science. It was partly as a result of this demeaning episode that in the following year, at the age of sixty-eight, Broom was finally given a junior post as a paleontologist at the Transvaal Museum, in Pretoria.

In 1925, shortly after the appearance in print of Dart's description of the Taung skull, Broom had visited Dart's laboratory in Johannesburg. The story goes that Broom burst into the room and, ignoring its occupants, strode over to the table on which the skull reposed, dropping to his knees in adoration.

Broom's examination of the skull, he told Dart, revealed nothing to contradict Dart's conclusions. In a published note he described the specimen as a Pleistocene link between ape and man and alluded to the probability that adult fossils would soon be found.

But Broom found himself virtually alone in supporting Dart. Dart retreated to the study of neuroanatomy, his first love, and paleoanthropology in South Africa languished until Broom was ensconced at the Transvaal Museum. At the museum he put the finishing touches to his work on the Karroo reptiles, then in 1936 decided to transfer his attention to the search for an "adult Taung ape."

By this time, however, the Taung deposit had been largely worked out, and besides, Broom could not afford to get there. So he turned his attention to sites in the more immediate vicinity of Pretoria. The deposits which at Taung had yielded *Australopithecus* had been cave-fill; and several similar sites in the Pretoria area were being worked in the same fashion. Indeed, the manager of a quarry at Sterkfontein, not far away, was the same man who had been in charge at Taung when Dart's discovery was made.

Broom got wind of fossils turning up at Sterkfontein and within months of the start of his search had recovered a partial *Australopithecus* skull at the site—this time an adult. The specimen, accompanied by another endocranial cast, was badly crushed and distorted but retained four teeth, and this was enough evidence for Broom to ascribe it to the genus *Australopithecus*. In the belief that its geological age was younger than that of Taung, however, he put it in its own species: *A. transvaalensis*. Subsequently Broom changed his mind and moved it to its own genus, as *Plesianthropus* ("near-man") *transvaalensis*.

Scientific reaction to Broom's discovery was disappointing. Keith's lengthy dismissal of the Taung baby as an ape in the 1930 edition of his book *New Discoveries Relating to the Antiquity of Man* was regarded as the definitive word on the specimen—and now, by extension, on anything else emanating from South Africa. A polite lack of enthusiasm on the part of his colleagues during a visit to Europe in 1937 failed to diminish Broom's zeal, however, and he redoubled his efforts on his return. By offering to pay for interesting fossils, Broom obtained several more fragments from Sterkfontein, and then, by great luck, a palate from a nearby cave site called Kromdraai. A visit to the site enabled Broom to assemble enough pieces to reconstruct the skull of the Kromdraai hominid. Clearly, it was different from its relative at Sterkfontein. Whereas the skull of the latter was relatively small and lightly built, this new hominid was larger, heavier, flatter-faced, and more robust. Broom's first reaction was that this bigger creature was yet more manlike than the Sterkfontein/Taung creature, and in 1938 he named it *Paranthropus* ("next to man") *robustus*. Again the scientific establishment was unreceptive. Even Broom's discovery of the lower end of a femur which showed evidence of upright posture left it unmoved. During the years of the Second World War, while the world's attention was considerably distracted from paleoanthropology, Broom continued work on blocks of matrix from Kromdraai and compiled a major comparative monograph on the South African material in which he concluded that all species belonged to the single subfamily Australopithecinae, that all were bipedal and probably tool-using, and that an as yet undiscovered Pliocene australopithecine was the ancestor of modern man.

Broom still had an all too real problem, however. And his problem lay not in the fossils themselves but in the sites from which they came. Most fossils are found in sedimentary rocks which form vertical sequences of strata running from oldest at the bottom to youngest at the top. Such sequences can be traced out to other sequences elsewhere, or can at least be compared to them on the basis of the fossil fauna they contain. This provides the basis for a system of relative dating of fossils vis-à-vis each other and for their allocation to the various epochs of the geological past.

The South African sites, however, are isolated from such sequences. The rocks which enclose the fossils represent consolidated rubble which accumulated within caves formed by the action of water in the soluble limestone surrounding them. Most cave sites which have yielded hominid fossils—Choukoutien is a case in point—show a small-scale layering of deposits which reflects the history of occupation—or otherwise—of the site. But the South African caves were never occupied by australopithecines. They were formed below the ground as solution cavities totally contained within the limestone. Openings to the land surface were later developments, as the action of water continued on the soluble dolomite. Once such openings had been made, material could fall at random from the surface into the caves, which then gradually filled up, over an indeterminate and not necessarily continuous period, with a jumble of bones, dirt, pebbles, portions of fallen cave roof, and so forth. Consolidation of this rubble by the cementing action of filtering minerals led to the permineralization of the bones and their encasement in a rock-hard and heterogeneous matrix known as breccia; and it was this which was being blasted away by the lime-miners when the fossils were discovered.

In any event, each cave has had an extremely complex history, and dating the filling of the cave on the basis of the fossil vertebrates it contains has been a highly approximate procedure, only possible at all since the Plio-Pleistocene mammal fauna of Africa has become better known from sedimentary sequences elsewhere.

C. K. Brain, currently director of the Transvaal Museum, has offered an ingenious explanation of how hominid fossils may have found their way into some, at least, of the caves. The areas of Sterkfontein and Kromdraai at the time of the australopithecines was probably about as dry as it is today, when a grassy landscape is only occasionally relieved by trees. Often such trees, particularly the larger ones, are found growing by fissures which collect water and are in the process of becoming, or opening into, limestone caverns. Leopards are known to drag their kills into the branches of trees; bones then fall to the ground as the victim is consumed or decomposes. If the tree grows at the entrance to a fissure, the bones easily find their way into any cavern beneath. Brain found among the collections of australopithecine remains a piece of skull vault with two closely spaced holes in it. Holes which in their spacing exactly matched that of the canine teeth in a leopard jaw. Circumstantial evidence, at best, for the hypothesis that some, at least, of the hominid fossils are the remains of leopard kills. But circumstantial evidence is all we are ever likely to have.

The difficulties of dating notwithstanding, the South African hominids found a much more receptive audience as the world returned to normal after 1945. At the same time that the eminent Harvard anthropologist Earnest Hooton was echoing the stance of the Old Guard in limerick form,

> Cried an angry she-ape from Transvaal
> "Though Old Doctor Broom had the gall
>   To christen me Plesi-
>   anthropus, it's easy
> To see I'm not human at all."

Wilfrid Le Gros Clark, professor of anatomy at Oxford, was reviewing Broom's newly published monograph favorably in the pages of *Nature,* and on a subsequent visit to South Africa found himself able to document their hominid status in some detail. It was Clark's support that finally tipped the scales in Broom's favor, but Broom himself, now over eighty, was still to make the clinching discoveries which were to assign to the outer wilderness the few voices still raised in opposition.

In 1947 Sterkfontein yielded not only a perfect cranium, lacking only teeth, of a gracile (lightly built) *Australopithecus,* and a lower jaw (which even the aged Sir Arthur Keith conceded showed the association of a manlike jaw with an apelike cranium) but also a pelvis and vertebral column, together with associated fragments of leg bone. These pieces showed beyond doubt that their owner had in life been an upright walker, and the size of the braincase of the undistorted cranium—about 480 cc., well up at the high end of the ape range, but only half the size of that of *Homo erectus*—showed that erect posture had unquestionably preceded brain enlargement in the evolution of man. There was some question, nevertheless, about *relative* brain size; despite the large teeth of the Sterkfontein hominid, its body, to judge from the size of its postcranial bones, was substantially smaller than that of an ape. The ancient but far from decrepit Broom was by now unstoppable. In 1948 he began work at another site a little more than a mile away from Sterkfontein. This site, Swartkrans, yielded up remains of more individuals of the robust type Broom had already described from Kromdraai as *Paranthropus*.

Dart, meanwhile, emboldened by Broom's success, took up again the pursuit of fossil man which he had abandoned so many years earlier. Before the war a student of his, later to become a distinguished paleoanthropologist in his own right, had found the skull of another fossil baboon at the cave of Makapansgat, the site of a famous battle some two hundred miles north of Johannesburg. The year 1947 found Dart excavating the site, and fragmentary hominid fossils began turning up immediately. Struck by the presence of carbon in the deposits, which he mistakenly attributed to the use of fire by his hominids, Dart named his finds *Australopithecus prometheus*.

As he worked, Dart became increasingly impressed with the large quantity of broken mammal bones he was finding. Could these have been the remains of australopithecine hunters? And might not some of the remains have been used as tools—of aggression and of food preparation—by these hunters? Indeed, without such weapons, how could such small, vulnerable, upright (and hence not swift) primates as *Australopithecus* have survived on the open savanna? Stone tools up to then had been singularly lacking in the cave sites. Such reasoning led Dart to the idea of an "osteodontokeratic" (bone, tooth, horn) culture wielded by a vicious, cannibalistic, as well as simply carnivorous, *Australopithecus*.

As we now know, there are far better explanations than this for the fossil accumulations at Makapansgat, but unfortunately Dart's ideas were quickly picked up and elaborated on by the journalist Robert Ardrey, in a series of best-selling books. As we have already suggested, Ardrey's basic premise—that man is descended from a vicious killer—and the corollary derived from it—that whatever nasty traits man shows today are simply a matter of inheritance—have achieved wide popularity largely because they implicitly absolve us from responsibility for man's inhumanity to man. But they are founded in fancy, not in fact.

Since Broom's death in 1951, at the age of eighty-four, work has continued at the surviving australopithecine sites (Taung was completely destroyed by mining operations). More material has been recovered, notably at Swartkrans, where in a separate section of the deposits hominid material came to light that did not resemble the robust material already found, and at Sterkfontein, which has very recently yielded a skull somewhat different from those found in presumptively earlier parts of the deposit. We also have a better idea today of the dating of the cave sites, although uncertainty remains. But back in the 1950s all that could be said, on the basis of faunal associations, was that some of the sites appeared to be younger than others and that the younger sites were those yielding the robust hominid.

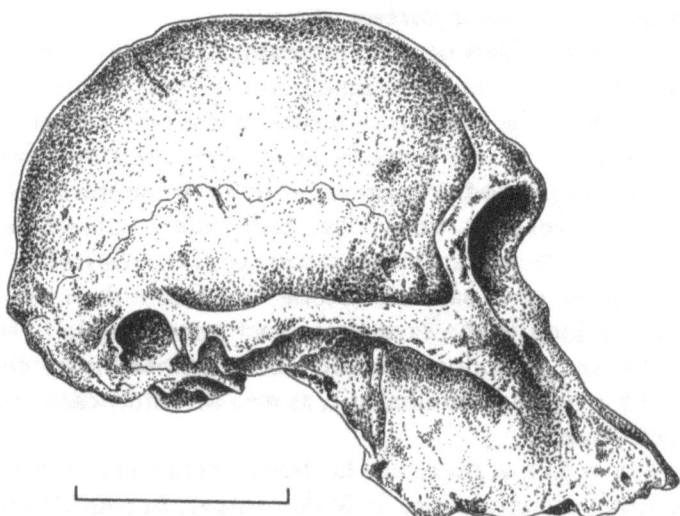

The best-preserved cranium of *Australopithecus africanus* from Sterkfontein, thought to be female. The scale represents five centimeters.

Absolute dates for any part of human evolution were still in the future; and when they actually arrived, they were to drop like a bombshell. Meanwhile, there gradually emerged a simplification of the nomenclature used to describe the early hominids from South Africa. At its peak, this involved some four genera and six species from the five sites known. This number was condensed by general agreement to one genus and two species: *Australopithecus africanus* for the "gracile" species from Taung, Sterkfontein, and Makapansgat; and *A. robustus* for the robust species from Swartkrans and Kromdraai. A more recent review has revived the concept of *Paranthropus* for the robust form; this seems to us to revive a useful distinction, but of that, more later.

Olduvai Gorge

Louis Leakey was one of the most remarkable characters in the history of paleoanthropology, and in several senses a fitting inheritor of the maverick tradition of Robert Broom. Born and raised in Kenya, Leakey was fascinated from childhood by the past, and in his early teens collected paleolithic implements near his home at Kabete. While an undergraduate at Cambridge, Leakey accompanied a British Museum fossil-hunting expedition to Tanganyika, and after graduation, while living a hand-to-mouth existence with no formal academic affiliation, managed to find funding for four archaeological expeditions to East Africa between 1926 and 1935. In 1931 he found the first stone tools at Olduvai Gorge, on the Serengeti plains in northern Tanganyika. This 300-foot deep canyon was eventually to prove the cornerstone of his career, but his investigations there almost led to an early downfall when Leakey mistakenly supported the great antiquity of a modern human skeleton which had been buried low in the deposits. During 1931 he also recovered several pieces of skull vault from a site to the west of Olduvai named Kanjera, and the front of a jaw from the nearby site of Kanam. Again, these finds led to Leakey's embarrassment when he was unable to substantiate the antiquity he claimed for the specimens.

Following these tribulations, Leakey moved permanently to Kenya where, with severely limited means, he searched extensively for early ape fossils as well as exploring the early history of man in East Africa. It was only in 1951, however, when a certain amount of private funding became available to him, that intensive work began at Olduvai. By this time numerous surveys had revealed the basic outlines of the Gorge's geology, and abundant stone tools had provided unequivocal evidence of the former association of early man with fossil faunas of increasingly archaic aspect. The Gorge is cut down through a series of almost-horizontal strata, and Leakey had early on divided the sequence thus revealed into five major beds, from Bed I, the lowest, with an early Pleistocene fauna, to Bed V, at the top, of quite recent age. Sedimentation of these deposits had not been continuous, but the Gorge clearly offered the potential of documenting man's evolution over a vast period of time. Already by 1951, Leakey had traced the archaeological record of Beds I—IV, running from the simple "Oldowan" pebble tools found at the bottom of the Gorge, to the more sophisticated stone implements of Bed IV, which he felt were comparable to the middle Pleistocene Acheulean hand-axe cultures of Europe.

Together with his wife, Mary, an accomplished archaeologist, Leakey threw himself into the intensive exploration of Olduvai. Apart from a couple of hominid teeth, however, the Gorge for years yielded nothing but rich archaeological and faunal records. Then one day in 1959, with funds virtually exhausted and Louis sick in camp with malaria, Mary found a hominid cranium at the locality, close to the bottom of the Gorge, which had yielded the first stone tools back in 1931.

Leakey received the find with mixed emotions. From almost the beginning of his career, and latterly almost alone in his profession, he had harbored the certainty that Africa would yield evidence of very early *Homo*. In his iconoclastic view *Australopithecus* was no more than a collateral branch of the human family tree, and Java, Peking, and Neanderthal men were no better; certainly, *Australopithecus* was incapable of making the stone tools found so abundantly at Olduvai.

Yet Mary's cranium was unquestionably that of an australopithecine. What's more, it was the cranium of a hyper-robust australopithecine, and by 1959 it had generally come to be considered that while gracile *Australopithecus* could legitimately be regarded as close to the ancestry of man, the robust form was no more than an aberrant terminal side branch. Where robust *Australopithecus* was flat-faced, heavily built, with evidence of enormous chewing muscles and huge molar teeth, this specimen outdid it in all these respects. The major chewing muscles on either side of the head met in the middle, to produce a large bony "sagittal crest" in the midline. The front teeth, the incisors and canines, were tiny, but the chewing teeth, the molars and even the premolars, were enormously expanded and flat-surfaced. The brain volume of this individual was estimated to have been about 530 cc, a little above the values from South Africa, but not much.

Leakey quickly swallowed his disappointment at this lack of substantiation of his preconceptions and named his new hominid *Zinjanthropus boisei*, after an old Arab word for East Africa, and his benefactor Charles Boise. Having thus separated it from the lowly *Australopithecus*, he could now recognize it as the toolmaker of Olduvai, and did so. In view of the undoubted stratigraphic association of the cranium with the crude Oldowan tools found at the same site (which, moreover, appeared to represent an actual hominid campsite—a living-floor—with the remains of butchered animals), few saw fit to contest the role of *Zinjanthropus* as toolmaker. Most authorities, however, rejected Leakey's separation of it from *Australopithecus*, claiming that it merely represented a new species of the latter.

The publicity that "Zinj" brought the Leakeys and Olduvai attracted funding that now enabled work at the gorge to proceed at an unprecedented pace. Rapidly these redoubled efforts paid off. In 1960 new hominid fossils began turning up. At the same locality as Zinj himself, low in Bed I, were found two broken

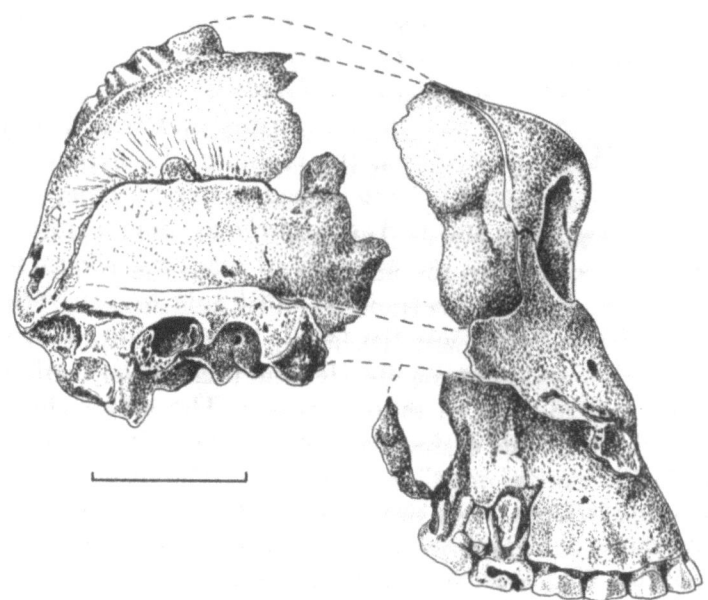

The cranium of the robust "Zinjanthropus" from Olduvai Gorge. The scale represents five centimeters.

lower-leg bones. From nearby, at almost the same stratigraphic level, and also on a living-floor, came some fragments of a hominid skull vault, some hand bones, some foot bones, and a lower jaw. Two adults were represented in this collection of fossils, and one juvenile. Leakey now saw his earlier beliefs vindicated, for the specimens were totally unlike Zinj. The jaw, for instance, was much smaller than the latter's must have been. Its molar teeth, while certainly large by modern standards, were much smaller, particularly relative to the front teeth. The bones of the cranial vault were lightly built, and reconstruction of the skull vault yielded a brain size which, at 680 cc., was noticeably larger than that of any *Australopithecus*. The leg bones and foot bones all said one thing: bipedal. After much thought, in 1964 Leakey and two colleagues published this form as *Homo habilis* ("handy man"), and included in the new species some specimens—teeth, jaw, and cranial fragments—from the middle part of Bed II. From being the maker of the Oldowan tools, Zinj became their victim—and that of their new maker, *Homo habilis.*

Moreover, Leakey saw evidence of the central place of his *Homo habilis* in human evolution when at the top of Bed II there turned up a cranium whose features—long, low vault of about 1,000 cc., thick-walled, with heavy browridges and sharply angled back end—clearly placed it with *Homo erectus* from Java and Peking. There was not the time, Leakey felt, to derive so different a form from *Homo habilis; Homo sapiens* must have descended from *habilis* without the benefit of passing through an *erectus* stage. This exactly echoed his declaration of 1932, when he had proposed that the broken skulls from Choukoutien represented the victims of meat-eating *Homo sapiens.*

Homo habilis was not cordially received in most quarters. Critics complained that the material was inadequate to justify so radical a diagnosis. The inclusion together of fossils from strata then thought to be as disparate in age as those from the bottom of Bed I and the middle of Bed II was widely held to be inadvisable, at the very least. Some said that the Bed I material represented *Australopithecus africanus*, and the Bed II stuff *Homo erectus*. And a major sticking point concerned the newly recognized absolute age of the former.

Shortly after its discovery, Leakey had announced his opinion that Zinj, faunally known to be from the lower Pleistocene, was in absolute terms about 600,000 years old. This was based on the prevailing estimate of around a million years for the duration of the Pleistocene as a whole, and Leakey's guess was generally regarded as reasonable, shot in the dark though it obviously was. Imagine, then, the astonishment when, in 1961, Leakey and two Berkeley geologists, Jack Evernden and Garniss Curtis, announced that the bottom of Bed I at Olduvai was about 1.8 million years old, and that Zinj was thus hardly younger. This radical new dating was by itself hard enough to adjust to; and now, approximately contemporaneous with Zinj, here was claimed to be *Homo*, at almost 2 million years ago.

The new dating at Olduvai marked the introduction to paleoanthropology of potassium-argon dating, a method which has had a profound effect on the field since its introduction two decades ago, and whose reliability has since been borne out by studies using other approaches to absolute dating. The technique depends on the fact that radioactive atoms possess unstable nuclei which spontaneously decay to stable states of lower energy. This they do at a constant rate, which is expressed in terms of their half-life, the time necessary for half of the atoms in a system to decay in this way. The half-life remains constant regardless of the number of atoms in the system.

One such atom is an isotope of potassium, ^{40}K, which is present as a constant proportion of newly formed potassium, and which decays to the rare gas argon (^{40}Ar) with a half-life of about 1.3 billion years. Measurement of the amount of ^{40}Ar which has accumulated in a sample of known potassium content thus enables one to calculate the time since the sample was formed. Of course, there are pitfalls. No argon must have been already present in the sample when the radioactive clock started ticking, and the system must have remained closed, i.e., it must have exchanged no argon with the outside. If argon was already present, the apparent age of the sample will be too old; if the sample has leaked argon, too young.

It is impossible to date fossils themselves using this method, but volcanic rocks are ideal for the purpose. First, they crystallize at very high temperatures at which no mineral can contain any argon, so argon present in the sample must have accumulated since the material last cooled. Second, volcanic rocks are precise stratigraphic indicators; in the geological succession a lava flow or a tuff (solidified ashfall) represents a single moment in time: the moment at which the radioactive clock started to tick. Since such structures usually interrupt a sequence of continuous sedimentary deposition, one can be fairly confident that fossils found in strata lying conformably above a tuff are slightly younger and that those contained in deposits just below are slightly older. The further one gets vertically from the dated layer, of course, the more approximate the extrapolated date becomes. From a practical point of view it is vital to be certain of two things: that a dated lava flow actually solidified on the ancient surface and was not simply extruded from below into a weakness between preexisting strata, and that samples collected for dating are uncontaminated by other material or have not lost any argon through weathering or other causes.

No such problem existed with the Olduvai samples. Repeated tests dated a tuff just above the bottom of Bed I at about 1.8 million years, a result since cross-confirmed by an independent method of radioactive dating. Meanwhile, knowledge was rapidly accumulating on other aspects of the geological history of the gorge, and this, combined with Mary Leakey's archaeological findings, was beginning to paint a picture of the life of early man less conjectural and more vivid than any that had been possible before.

During the time that Bed I was being deposited, the Olduvai basin was occupied by a lake several miles long. Neighboring volcanoes periodically spewed out ash which contributed not only to the deposits later reexposed in the walls of the Gorge but also to the alkalinity of the lake which, with the consequent concentration of algae, supported a rich fauna both in its waters and along its edges, where hominids camped. The climate, relatively wet when the early levels of Bed I were being laid down, became steadily drier through later levels (about 1.9–1.65 million years ago). Early Bed II sediments show recovering humidity, but by the end of Bed II times (around 1.2 million years ago) the lake had all but gone. Throughout this time, hominid occupation was focused upon the margins of the lake itself and, as it diminished and disappeared, on the banks of streams which drained into the Olduvai depression.

From earliest times at Olduvai, stone tools existed in variety at such campsites. The "Oldowan" stoneworking tradition did not consist merely of "crude chopping tools" made to an approximate pattern. Mary Leakey has been able to identify several recurring types, some of which, such as cleavers, scrapers, awls, and so forth, served obvious functional purposes. There is some refinement of this kit over time, as the tradition survived up through Beds I to IV, but the culture remained essentially the same. However, in middle Bed II times, as the climate changed and the lake finally dried up, there appeared, as if from nowhere, a new and more sophisticated tradition. This new stoneworking technology resembled the Acheulean "hand axe" cultures of Europe in that tools were fashioned from flakes struck off a prepared "core"; they were not, as in the Oldowan, cores shaped to a pattern or small flakes. The two industries survived at Olduvai, side by side, for hundreds of thousands of years.

The life-style of the hominids at Olduvai seems to have shown relatively little change, too. There is presumptive evidence from campsites that from the earliest times brush shelters were erected at these sites and that different activities—reflected in the nature of the bone and stone tool litter—were carried on inside and outside the shelters. At each site, broken animal bones attest to the fact that animals were brought "home" to be eaten, and there is some evidence that as time went by, larger animals were killed with increasing frequency (at a very early site the skeleton of an elephant was found in association with a large number of stone tools, obviously used to butcher it; but it seems possible that the hominids involved had merely butchered a carcass they found). Studies have also been made of the various rock types used to make the tools found at the different localities. In Beds I and II, which span a period of close to a million years, most tools were made of lavas found within a mile or two of the campsites; relatively few were made of material from farther away. In later levels, both the variety of rocks and the distance of the sources tended to increase, and it has even been suggested that trading in stone was carried on between early hominid groups.

Omo

With eastern Africa established by the Olduvai discoveries as the new focus of paleoanthropological attention, exploration of that part of the world by those interested in fossil man was bound to intensify. It was known that extensive Plio-Pleistocene sediments existed in northern Kenya and the contiguous area of southern Ethiopia, notably in the basin of the Omo River, which drains the Ethiopian highlands toward the south. The Omo deposits had been visited before the Second World War by a French paleontologist, Camille Arambourg, but interest subsequently lapsed until 1959, when Leakey encouraged an American anthropologist, Clark Howell, to undertake a survey in that area. Howell's brief reconnaissance revealed a thick sequence of deposits, many of them fossiliferous and with obvious paleoanthropological potential; but it was not for several years that firm plans were made to follow up the survey with more intensive work.

Several factors contributed to the delay, among them the difficulties involved in persuading the Ethiopian authorities to give permission for an expedition. As luck would have it, however, Leakey found himself introduced to Emperor Haile Selassie when the latter made a state visit to Kenya. When asked why it was that Kenya and Tanzania were rich in fossils while Ethiopa was not, Leakey responded that Ethiopian fossils indeed there were. They just had not been found, he explained, because of the problems of persuading the bureaucracy to give its blessing to the necessary activities. Within hours, the necessary permissions were available.

As a result, 1967 witnessed the formation of a joint French-American-Kenyan Omo Research Expedition. The aged Arambourg was in charge of the French contingent, Howell of the American. Louis Leakey would have represented Kenya, but ill-health intervened, and Leakey's son Richard was sent in his stead. Eight years of fieldwork in the Omo deposits led to the decipherment of a fairly complex geology and to the recovery of large numbers of mammal fossils; but the yield of hominid fossils was rather disappointing. Not that quite a lot of them weren't found; they were, and at very many different localities. But most of them were in the form of isolated teeth, and the number of more complete specimens was rather low. One of the problems, it seems, is that a large part of the Omo deposits was laid down by rather fast-flowing water, not highly conducive to the preservation of animal bones. And, of course, even under the most favorable of circumstances the different parts of the skeleton do not all have equal chances of being preserved. To put things in perspective, any vertebrate paleontologist knows from experience that by far the greatest number of fossils he finds, under almost any conditions of sedimentation, are single, isolated teeth. This is easily understandable, for dental enamel is the toughest substance in the body and hence the least susceptible to postmortem destruction. Unidentifiable bone fragments apart, the next most common sort of fossil is a bit of lower jaw with one or two teeth in it. Bits of (less solid) upper jaw are much less common yet; complete jaws are something of a rarity, and whole or even partial skulls are exceptional. Another factor to consider when assessing the entire faunal collection from an area is the original abundance of the animals represented in it. Hominids, it seems, were never abundant on the landscape, at least until very recently. Grazing mammals roamed the plains of ancient Omo in vast herds, and the fossils recovered reflect this. Hominids were thin on the ground, and the fossil record reflects this, too.

But reluctant to produce hominid fossils as Omo proved to be, it was wonderfully unique in one critical respect: the stratigraphic record. The pile of sediments was almost 3,000 feet thick, covering a huge span of time. What's more, tilting of the strata meant that successively older sediments were exposed by erosion at the ground surface; a 3,000-foot-deep gorge was not required to reveal the oldest layers. And to cap it all, numerous datable volcanic tuffs were interspersed at short intervals between the sedimentary layers. The process of unraveling the sequence in time of the localities at which the many thousands of fossils were recovered was no simple affair, but the upshot of it all was a datable sequence of faunas running in age between about 4 million and 1 million years. At last there was a kind of yardstick of Plio-Pleistocene time, calibrated both in closely bracketed potassium-argon ages and in the mammals characteristic of each age.

What of the hominids themselves? Less than spectacular though most of them were (with the exception of a couple of eye-openingly robust lower jaws), at the very least they extended the hominid fossil record well back into the Pliocene. Prolonged deliberation over his sample brought Howell to the conclusion that four kinds of hominid were present in it. Most abundant of the Omo hominids was an extremely robust type, with a super-massive lower jaw and huge back teeth like those of Zinj. This type was restricted to localities between about 1 million and 2 million years old. Older than this were teeth somewhat resembling those of *Australopithecus africanus*; these made their appearance at about 3 million years ago, and persisted for between half a million and a million years. A few teeth appeared to resemble those of *Homo habilis* from Olduvai (although this species is hard to diagnose on the basis of its teeth) and were of about the same age as the Bed I material from Olduvai. Finally, at the very top of the section, *Homo erectus*-like teeth began to turn up, in localities at about the 1.1-million-year mark.

This diversity of hominids, and their distribution in time, is certainly not without interest. Nonetheless, the most profound significance of Omo to date has lain in its dated faunal sequence. But one spin-off from the Omo Research Expedition has overshadowed even this. Read on.

East Turkana

When the Omo basin was divided up among the various national contingents of the Omo Research Expedition, the twenty-three-year-old Richard Leakey found himself obliged to examine his situation with care. Not only had he been assigned the youngest and least promising area of the deposits, but lacking formal training in paleoanthropology he feared that he would be overshadowed by the established scientists of the expedition. This was not a position he relished, and he cast about for an alternative. Making use of a helicopter hired by the project, he headed south into Kenya and overflew a series of exposures on the eastern side of Lake Turkana (then called Lake Rudolf) which had previously looked promising to him from the air. On landing he found paleontological riches beyond his wildest dreams: Plio-Pleistocene fossils eroding out of the sediments on all sides. Leakey abandoned the Ethiopian project, organized his own team, and turned his attention to East Turkana.

The first field season at East Turkana, in 1968, was devoted to survey work and confirmed the enormous fossil potential of the area. As a matter of policy few fossils were actually collected at this stage, but among the small collection which was made were four hominid fragments. The following year a more elaborate operation was mounted, with a base camp set up at Koobi Fora, a sandy spit sticking out into the lake. Early finds in 1969 included stone artifacts embedded in a volcanic tuff. This locality became known as the Kay Behrensmeyer Site, after its discoverer, and the tuff as the KBS tuff. Samples of this tuff sent for potassium-argon dating yielded an age of about 2.6 million years. Remarkably, of five hominid fossils collected in 1969, two were crania. One of these was the greater part of the braincase of a lightly built hominid, and the other was a virtually intact cranium of a robust australopithecine. This latter specimen was of special interest, since it answered a question which many had been asking about the Zinj skull from Olduvai.

One of the most remarkable differences of Zinj from the robust South African australopithecines was the great depth of its face. Since the discovery in 1964, near Lake Natron in northern Tanzania, of an almost complete lower jaw whose dentition closely matched that of Zinj but whose vertical ramus was much less high than would have been required of the matching lower jaw, anthropologists had wondered whether the great facial depth of Zinj was simply an individual variation and not characteristic of its species as a whole. Now this suspicion was confirmed. Like Zinj the new skull, ER-406, with a brain volume of a little over 500 cc., was enormously robust. But it was much shallower-faced. The specimen lacked teeth, but the roots preserved in the upper jaw clearly indicated a dentition like that of Zinj. In recognition of its "yet more so" differences from the robust South African hominids, the new specimen, together with Zinj, was assigned to the species *Australopithecus boisei*.

In subsequent years East Turkana hominid fossils accumulated at an impressive rate, and a large team of specialists was assembled to study the geology and archaeology of the area as well as the various kinds of fossil mammals recovered. In 1970 a variety of hominid jaws turned up, together with a partial skull which could have belonged only to a female *Australopithecus boisei*. This find finally settled a long-standing argument over whether the robust and gracile South African australopithecines actually represented male and female of the same species.

Most higher primates, and especially those which spend a lot of time on the ground, display sexual dimorphism, i.e., size and shape differences between males and females. The males are

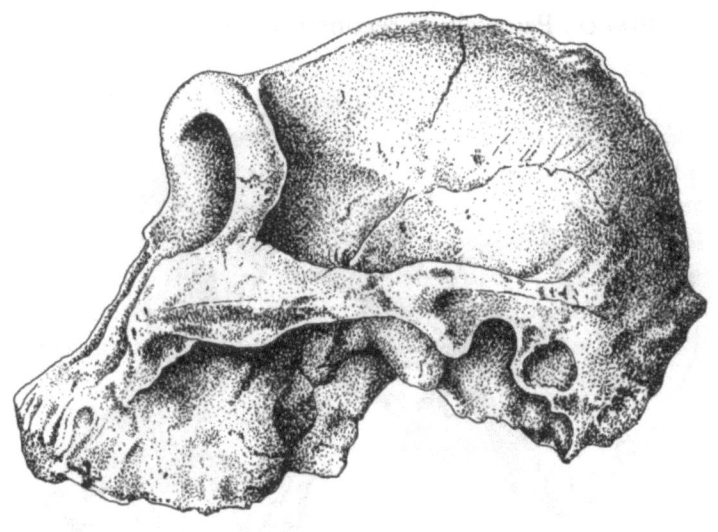

Robust *Australopithecus* from East Turkana: ER-406. The scale represents five centimeters, and the specimen is believed to be male.

larger and more robust and often have disproportionately larger canine teeth. Adherents to one school of thought had argued that in the past such differences between human males and females were larger than they are today and that this was the simplest explanation for the differences between the robust (putatively male) and gracile (female) australopithecines. Although this was never more than a minority viewpoint and was rendered inherently unlikely by the fact that the "males" and "females" were found separately in different sites of different ages, the proponents of this idea stuck tenaciously to it until ER-732, as the new cranium was known, showed that while "robust" females were certainly more lightly built than their mates, they nonetheless did not resemble gracile *Australopithecus*.

More jaws and postcranial bones were subsequently found at East Turkana, including jaws which looked remarkably like those of *Homo erectus*, and leg bones of equally modern aspect; but the next bombshell did not fall until 1972, when many pieces were found of a skull whose cranial capacity fell not far short of 800 cc. Painstaking reconstruction of the skull, known

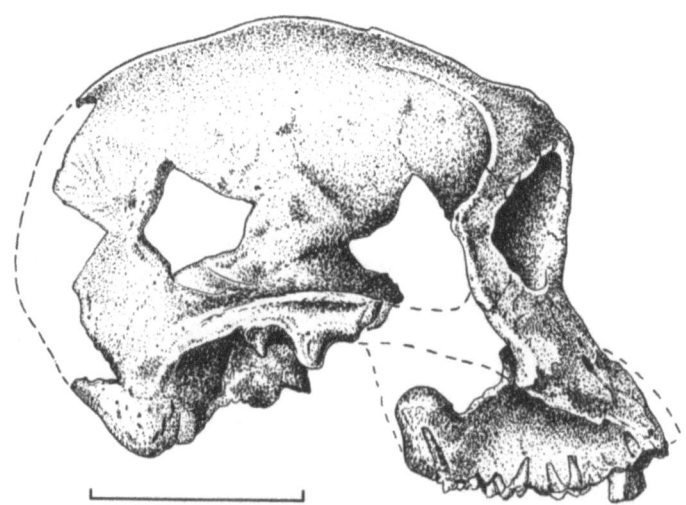

Presumed female robust *Australopithecus* from East Turkana: ER-732. The scale represents five centimeters.

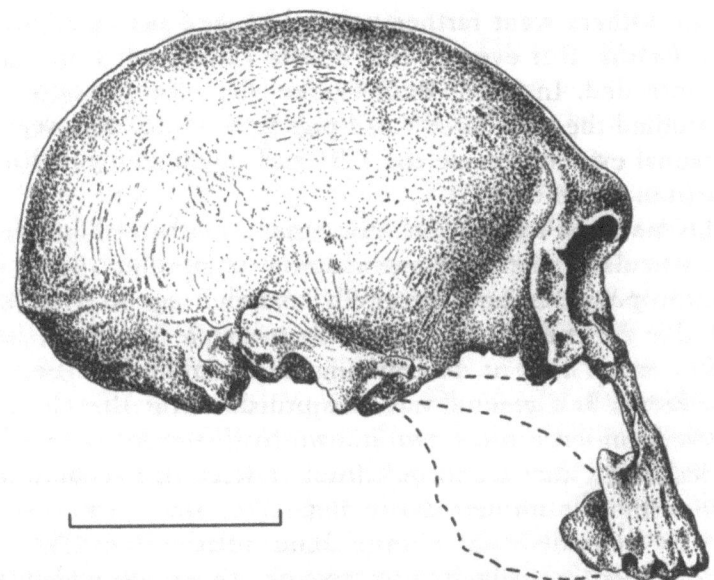

Homo habilis from East Turkana: ER-1470. The scale represents five centimeters.

as ER-1470, revealed a creature with a brain much larger than that of any known australopithecine, not robust at all but with a flatter face than *Australopithecus africanus*. To judge from what was left of the roots of its teeth, the back teeth were pretty large, but the front ones were not reduced as in *Australopithecus boisei*. Yet the real shocker lay not in the combination of ER-1470's large brain with any other feature of its anatomy, but with its claimed antiquity. For the deposits in which the fossil had been found lay below the KBS tuff, and the KBS tuff was dated at 2.6 million years. ER-1470 was therefore even older—maybe 2.9 million years, it was suggested.

What had Leakey discovered with ER-1470? Opinion was divided, even among those who worked most closely together on the Koobi Fora Research Project, as the East Turkana Expedition was by then known. One point of view held that ER-1470's affinities lay with *Australopithecus*; another, Richard Leakey's, that its large brain admitted it to *Homo*, species uncertain. Others went farther yet and viewed the specimen as *Homo habilis*. But even as this debate continued, a new sour note intruded. In 1973, Basil Cooke, the paleontologist who had studied the well-dated fossil pigs from Omo, declared that the faunal evidence made the KBS tuff much younger than its potassium-argon date.

This mattered a good deal: more than it perhaps should have, for particularly since the introduction of absolute dating paleoanthropologists have generally regarded age as absolutely central to the evaluation of fossils. In principle, however, dating doesn't much help in determining the relationships between fossil forms. In a general way, it is probably true that the more remote from each other two known fossil species are in time, the less likely they are to be closely related. But evolutionary relationships do not necessarily depend on dates, and certainly dates by themselves say nothing about relationships. What the dates of fossils actually do is to give one the *minimum* dates for the appearance of the species to which the fossils belong; in reality, it is only the analysis of morphology which can reveal the pattern of evolutionary relatedness among organisms. Nonetheless, since evolution is usually equated with change—constant change—and the evidence for and the implications of nonchange are widely ignored, dating has been regarded as being of the essence. If a form is early in time, the myth runs, it can, if there is nothing really remarkable in its morphology to prevent it, be conveniently plugged into the scheme as an ancestor. If it occurs late, then it qualifies with little more ado to be a descendant.

Thus the dating of ER-1470 appeared to be critical to the interpretation of its relationships and, by extension, its classification. Understandably under the circumstances, the first instinct of the East Turkana group was to close ranks around the earlier date. After all, ancestors are, on the face of it, more fascinating than mere descendants. However, dating of further specimens from the KBS tuff in the laboratory of Garniss Curtis at Berkeley consistently yielded a date of around 1.8 million years, and to cap this, another analysis of the East Turkana pigs, this time by members of the Koobi Fora group, came up with conclusions similar to Cooke's. So ER-1470 became about 1.9 million years old, which made it easier for most people to accept as *Homo habilis*, because it was now of about the same age as *habilis* from Olduvai.

The series of discoveries at East Turkana did not end with the finding of ER-1470. The year 1973 saw the discovery of two more skulls, both having many of the attributes of gracile *Australopithecus* and brain volumes of about 580 and 510 cc. respectively. Then, in 1975, a cranium completely different from any yet found in the area turned up: ER-3733, with a brain volume estimated at around 850 cc. and characteristics clearly aligning it with *Homo erectus*. This specimen was found in the Upper Member of the Koobi Fora Formation, at about the same level as the robust ER-406, and both are thus around the same age, close to 1½ million years, or perhaps a little more.

The time span covered by the bulk of the hominid fossils from East Turkana thus runs from around 2 million years to something under a million and a half. A few teeth have been recovered from older sediments underlying the Koobi Fora Formation, and more recent archaeological traces have been found, but the major record from East Turkana preserves a period of under a million years straddling the Pliocene-Pleistocene boundary. During this period it appears that the area was somewhat damper than the semidesert of today, and was partly occupied, as now, by a lake. Probably the lake, which fluctuated in extent and was fed by both seasonal and permanent rivers, was fringed by a lush vegetation which graded into savanna away from the water. The sites which have yielded fossils attributed to *Homo* seem to be concentrated along the ancient lake margins, while robust *Australopithecus* fossils are found equally on lake and river edges. Two stoneworking traditions have been identified at East Turkana: the KBS industry, which is equivalent both in time and technique to the Oldowan of Bed I at Olduvai; and the later Karari industry, which is dated to between 1.2 and 1.5 million years and which is more variable in tool types than is the KBS. The relationship between the

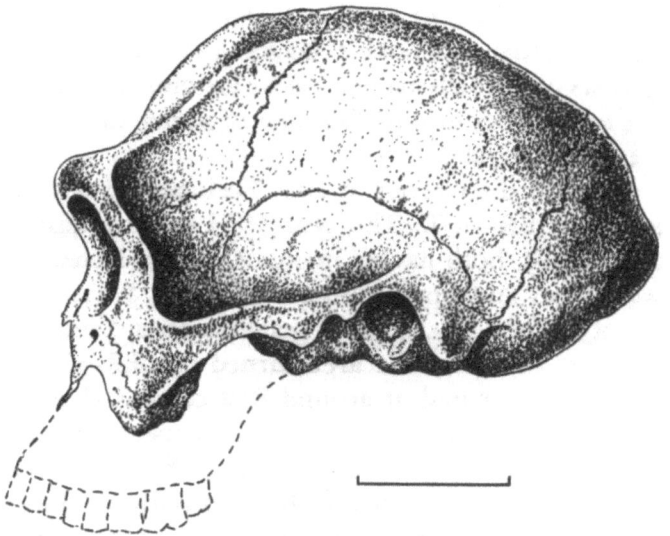

Homo erectus from East Turkana: ER-3733. The scale represents five centimeters.

two industries is not clear, and the archaeologists are being pretty cagey about it. Broadly, however, the Karari seems to equate with the Developed Oldowan of Olduvai.

Hadar

One of Clark Howell's field assistants in the Omo, Donald Johanson, was a graduate student who developed a close rapport with members of the French contingent. During a subsequent visit to France, Johanson encountered a French geologist, Maurice Taieb, who had been studying the geology of the Afar region of northern Ethiopia. Part of the Afar, Taieb told Johanson, consisted of badlands with fossils eroding out widely. He guessed them to be of Plio-Pleistocene age and suggested that Johanson join him on a visit to the area. The results of a preliminary survey in 1972 were so encouraging that Taieb and Johanson, together with Yves Coppens, who had taken over leadership of the French Omo contingent on Arambourg's death, agreed to organize a major international expedition to Hadar, as the most promising area of deposits was called. The strata at Hadar had been laid down on an outwash plain with high ground to its west. Streams drained into a small lake which grew as the climate grew wetter, then contracted once more as drying set in. Throughout the time of deposition local volcanic activity produced potentially datable ashfalls and basalt flows. Subsequently there was little faulting or tilting of the deposits, but these have been exposed in a labyrinth of erosion gullies. The mammalian fossils picked up during the initial survey seemed to suggest an age of around 3 million years.

The first major field season at Hadar, in 1973, produced the lower part of a femur and the upper part of a tibia (shin bone) of a small hominid. Not so startling, might be one's first reaction. But as it happens, the knee joint provides a telling piece of evidence about the kind of locomotion of its possessor. The apes and monkeys, for instance, are quadrupeds, supporting their weight on four legs. The stability of such a stance is improved by keeping the feet wide apart, so that the body's center of gravity falls well within the area defined by the supporting legs. Accordingly, the long axis of the leg of a quadruped goes straight down from the hip joint to the ground, and the axes of the major bones involved, the femur and the tibia, are in line.

A striding biped, on the other hand, has different requirements. Standing with the feet apart is one thing; but moving forward is another matter entirely. A chimpanzee walking bipedally has to move forward by swinging its legs out and ahead, swaying its center of gravity from side to side and using a large amount of energy to arrest this motion with each stride. People, however, can stand with their feet together, and when walking they move them forward in a straight line which is also followed, efficiently, by the body's center of gravity. This is achieved by a specialization in the bony structure of the leg. For obvious reasons, the hip joints are widely separated; but to compensate for this the long axis of the femur is directed obliquely down so that the knee joints fall close together in the body's midline.

The tibias then descend directly to the ground. The joint surface of the knee is parallel to the ground, so the long axis of the femur is oblique to that surface. The angle thus formed is known as the "carrying angle," and is diagnostic of human-style bipedality.

Small though the Hadar knee joint was, it had a perfect carrying angle, and was uncontestably the knee joint of an upright biped: a hominid. Close by, two upper femur ends were found, possibly belonging to the same individual. Before leaving Ethiopia, Johanson announced the find at a press conference, estimating its age at between 3 million and 4 million years and thus at a stroke almost doubling the antiquity of upright walking among hominids. The date was a guess based on the comparison of associated fossil mammals with the calibrated Omo sequence; basalt samples from a lava flow in the sequence had been sent for potassium-argon dating, but the results were not yet available.

The 1974 field season started with a bang. Within a couple of weeks of the beginning of prospecting several hominid jaws were found, including a complete palate with all the teeth in place. These dentitions were rather different from anything found anywhere before, and the size disparity between the smallest and largest specimens was remarkable. More was soon to come, however. Shortly afterward, "Lucy" turned up. Lucy, as all the world now knows, is the skeleton (or at least, a lower jaw, some skull fragments, most of both arms and a leg—no hands or feet—and parts of the spine and rib cage) of a tiny young adult female hominid. The same year also saw progress on some geological problems that had persisted at Hadar. A basalt flow that had been thought to lie relatively high in the stratigraphic section was relocated lower down and was assigned a potassium-argon date of at least 3 million years. The jaws and the previous year's knee joint had been found in strata well below the level of the basalt, and were thus earlier in time; Lucy was somewhat later.

The year 1974 had been one of great political turbulence in Ethiopia, and the following year was even more fraught with uncertainty. Fieldwork in the Afar continued, however, and paid off once again in spades. This time a locality incredibly rich in fossil hominids was found, above the level of the basalt flow but lower in the section than Lucy. Altogether this astonishing patch of sediment yielded the remains of at least thirteen individuals, adults and children alike, and represented almost 200 separate pieces. Perhaps more remarkable yet was that this pocket produced hominids alone; the mammals found so abundantly elsewhere at Hadar were totally absent. Such a situation was unprecedented. One suggestion was that the First Family, as the assemblage was fondly dubbed, was the remains of a band of hominids trapped in an arroyo by a flash flood. Unprovable, but as good a possibility as any. In any event, however nature had actually assembled it, this group of fossils provided a unique cross-section of early fossil hominids from a single point in time.

The year 1975 also saw the collection for dating of samples from a volcanic tuff high in the section of Hadar. These samples were dated both by potassium-argon and by another method known as fission-track, which involves counting the scars left in the structure of crystals by the explosive decay of uranium-238 to lead. Both methods yielded an age of about 2.6 million years. More recent work has raised this date to about 2.8 million, and the basalt layer earlier given a minimum age of 3 million years has since been redated, with little room for error, at 3.75 million. This redating is, moreover, in line with what faunal studies were suggesting for the age of the deposits, again by comparison with the Omo sequence. The time spanned by the fossil hominids of Hadar thus appears to be on the order of a million years, with the original knee joint and the 1974 jaws at about 4 million, Lucy at something over 3 million, and the First Family in between, at maybe 3½ million.

Each field season at Hadar had produced something remarkable, and the 1976/77 season, which was the last for several years, continued the tradition. At the end of 1976 the earliest stone tools ever found turned up, in sediments dated to around 2.5 million years, but so far devoid of hominid fossils. This date makes the implements fully a half million years older than any known from elsewhere. Not many of these tools have so far been recovered at Hadar, but those which are known are not apparently greatly dissimilar from those found in the lowest levels at Olduvai.

Laetoli

Back in 1935, Louis Leakey had picked up some fossils at a site called Laetoli, a couple of dozen miles south of Olduvai. One of these was a hominid canine, but it was misidentified and Leakey turned his attention to Olduvai without realizing what he had found. He was followed at Laetoli by the German paleontologist Kohl-Larsen, who found a bit of hominid upper jaw, but whose excavations were then terminated by the outbreak of war in 1939. Apart from a couple of brief visits the site then lay fallow until 1974, when Mary Leakey recommenced operations there. Immediately she began finding hominids, although nothing on the scale of what was simultaneously coming to light from Hadar. The hominid fossils consisted mostly of single teeth, but a partial juvenile lower jaw was found and an adult lower jaw with more than half its teeth still present. These specimens came from sediments lying between two volcanic tuffs radiometrically dated at about 3.8 million and 3.6 million years respectively.

But the real superstars of Laetoli did not turn up until 1977. The previous year it had been noticed that some exposed tuff surfaces near the camp bore impressions which appeared to be animal tracks. Apparently, some 3.6 million years ago a nearby volcano had puffed out a cloud of fine ash which had covered the landscape with a thin layer of powdery cinders. Rain then fell, turning the ash into something which must have resembled wet plaster of paris and which faithfully recorded the footprints of the animals walking across it. These tracks subsequently dried out and hardened in the sun, and were then covered with new ashfalls which preserved them intact. Among the animals which recorded their passage in the wet ash were hominids, and in 1977 their footprints were recognized. There was a good deal of argument at first about the tracks and what they represented: they were blurry, and there were not many of them. But in 1978 more hominid footprints were uncovered, and in 1979 yet more. There was no longer any question what they were.

These footprints have a very special importance. Although paleoanthropologists had heatedly debated for many years over whether upright posture had preceded or succeeded the large brain in human evolution, it has for a long time been clear that upright posture has characterized man's forebears from very early times. But while upright posture among early hominids has been almost universally accepted in principle for many years, there has been a widespread underlying belief that early hominids were somehow less efficient bipeds than we are. Think of Boule and Neanderthal man, for instance. Even recent studies show this tendency; thus analyses of the Olduvai Bed I leg and foot bones concluded that although their possessors had been bipeds, they were less good at bipedalism than we are.

This expectation is, of course, as good an example as one is likely to find of what happens when evolution is viewed as inexorable progression. The explicit idea that evolution is directional, working toward a goal, was laid to rest before the end of the last century. But when the course of human evolution is viewed, so to speak, through the wrong end of the telescope, and man is regarded as some sort of pinnacle toward which earlier hominids were constantly striving, the results in practice are not dissimilar.

This is particularly true when the expectation of gradual change is added in, as it inevitably is. There is then a tendency to see fossils as forms in transition from one state to another, instead of as organisms existing in a particular set of environmental conditions. Le Gros Clark, that fine anatomist and defender of the fossil record, was led, for example, to expect that in the course of human evolution adaptation to bipedal locomotion started in the foot, and then over a very long period of time worked its way up to the hip via the ankle and then the knee. Despite the fact that it is extremely difficult to visualize a primate with a human foot and an apelike hip surviving successfully in the environments typical of humans, or of apes, or indeed in any conceivable intermediate environment, such ideas (and more insidiously, the view of evolution they embody) colored interpretation of hominid fossils for many years. Indeed, this view of the evolutionary process still does. It is only recently that Owen Lovejoy and his associates have suggested that early hominids may actually have been better bipeds than we are; that in acquiring the wide birth canal necessary to permit the passage of large-brained babies we may have had to compromise the efficiency and structural integrity of our locomotor apparatus.

The Laetoli footprints pushed aside for good the arguments for the inefficiency of early hominid bipedalism based on bony structure. Although the blurry 1977 prints allowed those with the kind of preconceptions we have been discussing to detect imperfections of locomotion in the gait of the hominids who made them, the clearer prints found later on made it absolutely evident that, whatever the details of its unknown anatomy (and the impression of the foot itself was similar to that of a modern foot), the creature actually walked just as we do. For once, function was directly observable and did not need to be inferred from form.

Ape-Man of the Afar

In early publications on the hominid material from Hadar, Johanson indicated that two species were present in the sample. Specifically, he expressed the view that while most of the specimens were allocable to a species of *Homo*, some, notably Lucy, were more closely allied with *Australopithecus africanus*. At the same time, Mary Leakey and her coworkers were publishing the view that despite their great age and their possession of certain primitive traits, the Laetoli mandibles and teeth should be placed in *Homo*. More recently, however, in collaboration with Tim White, the primary describer of the Laetoli hominids, Johanson has expressed the view that only a single hominid species is represented in the entire assemblage of fossils from both sites. His conversion to this viewpoint by White is entertainingly described in the recent book *Lucy*.

Johanson and White believe now that this single hominid is distinct from any previously described and have created for it the new species *Australopithecus afarensis*. The species is said to be distinguished by a suite of characteristics including large incisor teeth; oval front premolars oblique to the dental row; a gap between the upper incisors and canine, and lower canine and front premolar; jutting of the upper jaw beneath the nose; and a host of others. *Australopithecus afarensis* is also earlier in time—a million years earlier—than any other Plio-Pleistocene hominid species; only a few scraps, notably the lower end of a humerus (upper arm bone) from Kanapoi (about 4 million years old), and a piece of lower jaw with one tooth from Lothagam (about 5.5 million years) are older, and neither can be allocated with any confidence to a particular hominid species. Johanson and White have thus boldly concluded that their *Australopithecus afarensis* is a "stem" species: that it is the common ancestor on the one hand of *Homo* (giving rise to the sequence *Homo habilis* → *Homo erectus* → *Homo sapiens*), and on the other hand of *Australopithecus africanus*, which they view as having given rise subsequently to *Australopithecus robustus* (including *A. boisei*).

This scheme has not gone uncriticized. One claim is that Johanson was right the first time and that more than one species occurs at Hadar. Another is that the whole assemblage is not separable from *Australopithecus africanus*. And if all the material does belong to one species, is it related as claimed to the other hominids? Such questions are the province of the next chapter.

CHAPTER SIX

Patterns in Human Evolution

PERHAPS THE greatest single obstacle that exists to the recognition of pattern in human evolution is the matter of recognizing species in the fossil record. For the paleontologist, the practical problem boils down to one of morphological variation. Individuals within a species are not all identical, but individuals belonging to the same species tend to resemble each other more closely than they resemble individuals of other species. However, individuals resemble each other because they belong to the same species, while the paleontologist has to start from the other end and assign fossils to species on the basis of their physical similarity. As we have already seen, speciation can take place without much morphological differentiation, or it can fail to take place in the face of a great deal of it. So the crucial question, How much similarity is required to place an assortment of fossils into the same species? is not at all an easy one to answer.

One instance that springs readily to mind is that of lions and tigers. None of us has any problem telling these beasts apart when confronted with the living, breathing animals—or even stuffed ones in a museum. But even those most familiar with the skeletons of big cats find it difficult or impossible to discriminate between them with only the bones to go on. And bones are all the paleontologist ever has. In general, our feeling is that species tend to go unrecognized in the fossil record; and a consequence of that tendency, it seems to us, has been an overemphasis on the homogeneity of that record. This, of course, has supported the classic myth of the gradual and progressive nature of the evolutionary process.

How Many Species?

In view of all this it is hardly surprising that the major scenarios of human evolution that were espoused during the 1960s and 1970s were often identified by the maximum number of hominid species they allowed to coexist at any one point in Plio-Pleistocene time. The simplest of these views (at least in its logical structure) was the single-species hypothesis. The single-species adherents traced their intellectual ancestry back to Schwalbe's paper of 1906, in which the sage had suggested that modern man stemmed from Java man via the Neanderthals. Their rationale lay in the supposition that the acquisition of material culture by the earliest hominids had so broadened their ecological niche that it was impossible to conceive of the coexistence of two toolmaking creatures. Modern man is alone in the world; so, by extension, were his ancestors. However, perhaps the most fundamental influence underlying the single-species stance was the adherence by its proponents to a highly literal form of the great evolutionary myth of slow, gradual, and progressive change. So bewitched were the single-species people by the simple linear elegance of this myth that they were reluctant to see it marred by messy speciations.

As long as the single-species scheme had to accomodate only those hominid fossils discovered up through the 1950s, a reasonably respectable case could be made for it. Even though the South African australopithecine sites had yielded unquestionable evidence of two types of hominid, one possible interpretation of these types was that the large, robust forms were males and the gracile ones females of the same species. This supposition was justified by the conjecture that early hominids were strongly sexually dimorphic; in other words, that size and shape differences between males and females were large.

Modern man, though sexually dimorphic, is not very markedly so; but proponents of the single-species idea favored the proposition that a better standard of reference was furnished by the great apes, particularly the gorilla. Male gorillas are considerably larger than females, and many of their physical differences from female gorillas are related to this size difference. Single-species fanciers were fond of pointing out, for example, that male gorillas (like robust australopithecines) have a sagittal crest (a bony ridge running along the midline of the top of the head, onto which the large temporal chewing muscles extend), while the smaller, more lightly built female gorillas (like the gracile australopithecines) do not. This reasoning didn't cut much ice with opponents of the single-species theory, who pointed out that the two types were found at different sites. It was unlikely, they felt, that females would have gone to Sterkfontein to die, leaving the grief-stricken males to stampede across the valley to Swartkrans and become fossilized three-quarters of a million years later.

But even though it was pretty clear-cut that the segregation by site of the australopithecine types argued convincingly against their inclusion in the same species, the single-species hypothesis was bolstered by the fact that not all of the South African cave sites were of the same age: both of the sites which had produced robust fossils were younger than all of the sites with gracile fossils. What could be simpler and more satisfying than an evolutionary sequence leading from gracile *Australopithecus* through robust *Australopithecus* to *Homo erectus* and thence via the Neanderthals to modern man?

Now, there is indeed beauty in simplicity, but it is a meretricious sort of beauty if it is based on oversimplification. The problem with schemes of this sort is that they assume that each stage is a direct descendant of the one before it and the ancestor of the one following. These are dangerous assumptions. Essentially, approaching things in this way involves allowing dates to determine relationships, whereas, as we have already seen, there is actually no necessary connection between the two. And indeed, if we look at the morphology of the various stages in the sequence we find that the simple linear model will not hold up.

However, fossils have a special magic of their own that comparative studies just do not possess, and devotees of the single-species idea held onto it in its purest and simplest form until absolutely incontrovertible fossil evidence began to turn up that more than one fossil hominid species had indeed coexisted during the Plio-Pleistocene. Even the discovery of Zinj and Bed I *Homo habilis* in almost coeval levels at Olduvai did not deter those who claimed that in the evolving hominid lineage an "*Australopithecus* stage" (which included the latter) had given rise to a "*Paranthropus* stage" (which included the former). This was only possible by exaggerating the significance of the very small stratigraphic difference separating Zinj from "Pre-Zinj," the name sometimes informally given to Bed I *habilis*.

Thus it was only when at East Turkana incontrovertibly male and female robust *Australopithecus* were found in the same stratigraphic levels as lightly built forms assigned to *Homo* that it began to look as if the single-species people would finally be obliged to complicate their elegant single lineage with a branching event which put robust *Australopithecus* out on a limb. The clincher, of course, came with the discovery of a typical *Homo erectus* contemporaneous with robust fossils. Obviously, demonstration of this coexistence effectively destroyed the basic tenet of the single-species hypothesis that the hominid ecological niche is too broad to permit any other state of affairs. Likewise, the corollary that toolmaking was the original and basic hominid adaptation was put to rest by the realization that upright locomotion preceded the appearance of tools by a million years or more. Nevertheless, the underlying desire to see human evolution in terms of the simplest progression possible still lingers, and albeit in modified form the linear model is still cherished by a few.

The end of the spectrum of Anglo-American paleoanthropological opinion opposite to that occupied by the single-species enthusiasts is that of Louis Leakey and his intellectual descendants. We might call them the "three-species" group. Leakey, of course, had for many years harbored the notion that very early *Homo* was to be found in Africa. And when the gracile hominids from lower Beds I and II at Olduvai turned up, he believed he had clinched the matter. Leakey saw at Olduvai evidence for three lineages: a robust australopithecine represented by Zinj, *Homo erectus* in the form of the cranial vault from the top of Bed II, and *Homo habilis*. These added up in Leakey's view to evidence for three lineages, because for him *Homo erectus* lay off the path to modern man, who, he felt, is a direct descendant instead of *Homo habilis*. A question that bothered many at the time was the extent to which the *Homo habilis* fossils could actually be distinguished from South African gracile *Australopithecus*. Much of the gracile Olduvai material was fragmentary, and the Bed I type fossils, on which the name was actually based, did not differ much from gracile *Australopithecus*. In fact, the large brain size suggested by the rather long skull bones was the only one of the supposed distinguishing characters that was at all convincing.

This problem was solved to the satisfaction of several leading scholars by allocating the Bed I *habilis* fossils to *Australopithecus africanus*, and the Bed II specimens, which were thought to be very substantially younger, to *Homo erectus*. During the 1970s, however, two things occurred that appeared to many to legitimize *Homo habilis*. First, new dates placed Bed I and Bed II much closer in time than had generally been believed. Association of both sets of fossils in the same species was thus more acceptable to the gradualist outlook of most paleoanthropologists. Second, and more important, the new finds at East Turkana appeared to bolster Louis Leakey's concept of *Homo habilis*, particularly after the KBS tuff turned out to be younger than had been believed and the ER-1470 skull was revealed to be roughly contemporaneous with Bed I *habilis*. Here was another hominid, and a relatively complete one at that, that occurred at around the 2-million year mark and had a large brain completely outside the range of any *Australopithecus* from South Africa. Although Richard Leakey was at least initially disinclined to allocate ER-1470 to a particular species within genus *Homo*, others were not, and *Homo habilis* of Olduvai/Turkana began to gain considerable currency. Indeed, the species began to accumulate fossil members from elsewhere. An increasingly sophisticated understanding of the South African cave sites led to the realization that their histories had been complex; and along with various other fragments a couple of skulls that possessed relatively voluminous brain cavities seemed plausibly to belong with East African *Homo habilis*. These specimens came from Swartkrans and from a late part of the deposit at Sterkfontein, and both appear to be around 1.5 million years old. Similarly, some of the Javanese hominids were recognized to be of considerably greater antiquity than Dubois' original *Pithecanthropus erectus;* and this, added to a suggestion first made in the 1960s that they resembled Bed II *habilis* from Olduvai, seemed to make it likely that *Homo habilis* had been present in eastern Asia also.

The new understanding of the South African cave sites also showed that *Australopithecus africanus* from Taung, Sterkfontein, and Makapansgat was rather older than had been thought, with faunal dates spread between 2 million and 3 million years. The realization that this species went back rather farther in time than any definite *Homo habilis* brought the three-species idea closer to the "two-species hypothesis," which had held the middle ground of paleoanthropological opinion. This hypothesis was essentially an extension of the consensus view that had prevailed before the Olduvai discoveries. In this view, early hominids were of two types, represented by the gracile and robust australopithecines. The two, it was believed, had descended from a common ancestor along divergent adaptive paths. The robust form, with its huge, flat, grinding molar teeth, had specialized in vegetarianism; the gracile one, retaining a more generalized dental condition with some cutting edges, had become a hunting and scavenging omnivore. According to taste, those who held to the two-species view could believe that the cunning, tool-wielding gracile form had directly exterminated the robust one, in a manner perhaps suggested by the Kubrick-Clarke scenario, or that, more indirectly, the former had merely outcompeted the latter for the resources offered by the early

Pleistocene African savanna. In either event, the robust hominid became extinct, while the gracile one went on to greater glories which culminated in ourselves.

The two-species view was a highly adaptable sort of hypothesis, capable of absorbing a good deal of new evidence without essential change. There was nothing at Olduvai, for instance, that directly contradicted it. Bed I *habilis* was easily absorbed into the gracile lineage, and Zinj could be regarded as an East African variant of the robust. If Bed II *habilis* was then transferred to *Homo erectus*, the sequence of events was clear. It was not until East Turkana produced the large-brained ER-1470 at a staggering claimed antiquity that it seemed that the two-species idea might not be able to accomodate all comers, after all.

Perhaps the greatest appeal of the two-species hypothesis had been that, rather in the manner of the Established Church of England, it was able to accomodate within it a wide variety of opinion. There was, for instance, much internecine debate over what the two forms which all recognized were to be called. One preference was to call the doomed robust form *Paranthropus* and to place the gracile one in *Homo* in recognition of what it was to become. Most paleoanthropologists, though, opted for two species of *Australopithecus*: *A. africanus* and *A. robustus*. A recent review has revived the name *Paranthropus robustus* for the latter; we think that this makes a useful distinction, and we would further place Zinj and the Turkana robusts in their own species, *Paranthropus boisei*.

A major effect of the discoveries at East Turkana, particularly as they were initially misdated, was to raise doubts in the collective paleoanthropological consciousness about the workability of any of these major scenarios of human evolution, at least in their pristine forms. But for reasons obvious in retrospect, a compelling alternative was lacking. The subsequent discoveries at Hadar and Laetoli at first did little to fill the gap, especially since the Hadar deposits, at least, appeared to be of about the same age as the older East Turkana strata, while yielding dissimilar hominids. It is a tribute to the mesmerizing power of dates that most of the Hadar material was at first called *Homo,* thus putting it right in with the gracile Koobi Fora fossils.

Once the confusion over the dating of the East Turkana hominids was cleared up, then, the time was ripe for Johanson and White to come up with *Australopithecus afarensis.* This added a species to the roster, but at the same time offered something to everyone of all shades of opinion. For as Matt Cartmill has pointed out, the Johanson-White conception of *afarensis* as a "stem" species that gave rise to two distinct lineages made obeisance to each of the major schools of thought. The single-species enthusiasts, in this view, were right to insist that sexual dimorphism was great in early hominids (for *afarensis* embraces a great deal of size variation) and were also correct in believing that *Australopithecus* gave rise to *Paranthropus.* The two-species people were wrong only in making *Australopithecus africanus* antecedent to *Homo.* And the belief of the three-species adherents in the reality of *Homo habilis* and in the ancient origin of *Homo* was vindicated.

So much for the various ways in which the hominid fossil record, or at least its early stages, has been viewed in recent years. Quite clearly, the history of palaeoanthropology has not involved a search for pattern, but has, rather, involved the attempt to define species and to link them up in some way that satisfies their distribution in time. The time requirement—of linking up hominid species into time series—more or less set the limits within which relationships could be recognized, while preconceptions (for example, the belief in ancient *Homo,* or the predilection for the simplest possible linear model) appear to have had no small influence on the already inevitably subjective procedure of sorting the fossils into species.

Now, these traditional concerns of paleoanthropology are quite legitimate and proper; but the way in which they have generally been pursued points up one of the most pervasive myths in all of paleontology. This is the myth that the evolutionary histories of living beings are essentially a matter of discovery. Uncertainties in our interpretations of the fossil record are ascribed to the incompleteness of that record. Find enough fossils, it is believed, and the course of evolution will somehow be revealed. But if this were really so, one could confidently expect that as more hominid fossils were found the story of human evolution would become clearer. Whereas if anything, the opposite has occurred. The "discovery" myth, as it happens, is a more or less direct extension of the "progressive change" myth. If the history of an evolving lineage is one of slow, steady change, then each generation could be viewed as a link in a chain which might, at least in theory, be fully documented by fossil evidence. As we have pointed out, however, this scheme does not allow for the diversification of life, which we know has occurred. And if this diversification has taken place through the multiplication of species through speciation, then the fossil history of life is something that cannot be directly discovered. Instead, what we have is a diversity of species in time and space, the relationships among which have to be analyzed. Only if we eliminate the possibility of diversification, as for instance the single-species people did, can we assume that known early members of an evolving group are ancestral and that later ones are their descendants. Our lives would be greatly simplified if we could just draw lines on a time chart to join up earlier fossils with later ones in a progressive sequence. Unfortunately, we can't.

But nobody would deny that all living species must have had ancestors and that not all fossil species became extinct without issue. Descendants there are, and ancestors there must have been. The problem is, how do we recognize them? Of course, time is important in the sense that a later species cannot give rise to an earlier one. But species, as we have seen, can be long-lived, and the fossil record will ever only give us the *minimum* time-span occupied by a given species. Essentially, then, recognizing ancestors amounts to a matter of elimination: we know that an earlier species cannot be ancestral to a later one if it possesses evolutionary novelties which are lacking in the other. But even if a species is not disqualified from ancestry in this way we can never *know* if the relationship was actually of this kind. The problem with statements about ancestry is that they are not susceptible to disproof, and are thus essentially nonscientific. The only way we can make scientific (i.e., testable) statements about fossil species is to restrict ourselves to discussing relationship without specifying whether the relationship concerned is that between an ancestor and its descendant, or that between two species descended from a common ancestor. We are able to make testable statements of this more general kind because, after all, evolution does involve change, and relationship is reflected in the common possession of evolutionary novelties.

Now, we do not dispute that if we were to eliminate speculations about ancestry and descent we would be cutting out one of the most exciting and interesting aspects of paleontology. And we would not advocate that for a single moment. But we do caution that statements that go beyond theories of generalized relationship are just that—speculations. The problem with most analyses of the hominid fossil record is that they are conducted on this speculative level, where time is often allowed to dictate ideas of ancestry and descent without relationships at the more fundamental level having been first determined. To reanalyze the hominid fossil record in the way we believe is necessary is a massive task, and one which has barely been begun. But we can nonetheless look at the fossil evidence for human evolution in a general way to see whether the pattern we find fits with what we see elsewhere in the fossil record of life. This is what we will do in the next section.

The Hominid Fossil Record

Perhaps the most straightforward way of looking at the human fossil record is to take it chronologically, looking at the fossils in broad time-bands. This is made easier, at least at early stages, by the gaps that still exist in the hominid fossil record—those same gaps that were used to chop up the mythical evolving continuum. The earliest fossil specimens we have mentioned so far are those allocated to *Australopithecus afarensis*, and we will look again at these undoubted hominids in a moment. But before we do that, let us briefly glance yet further into the past, into the shadowy area covering the actual differentiation of the hominid line from that of our close relatives the apes.

During the Miocene, the geological epoch preceding the Pliocene and spanning the period between about 25 million and 5 million years ago, early apes were widespread over the Old World. Apes from the early Miocene were relatively unspecialized, although clearly apes; but until a few years ago it was believed by many that the hominid family could be extended back to the middle Miocene, some 11 million to 16 million years ago, when it was represented by a form known as *Ramapithecus*. Although *Ramapithecus* was known only from a few jaws and teeth, it was thought to be rather human in a variety of respects, including the shape of its dental arcade, the relative flatness of its face, and the reduction of its anterior teeth, particularly the canines. It was even thought in some quarters that this last characteristic made it probable that *Ramapithecus* wielded tools, for how else could it have defended itself with its inbuilt biological weapon now gone?

More recent researches, together with more and somewhat better fossils, suggest that this picture was rather oversimplified. It is now realized that a group of middle to late Miocene hominoids (members of the superfamily Hominoidea, which embraces both man and the apes), including *Ramapithecus*, *Sivapithecus*, and *Gigantopithecus*, possess dental specializations otherwise known among primates only in Plio-Pleistocene hominids. Among these features are thick dental enamel, large molar teeth as compared with inferred body size, at least somewhat reduced incisor and canine teeth, and strongly built jaws. Unfortunately the group is still known almost entirely from jaws, teeth, and facial fragments, so no very complete picture of its members can be drawn. However, it does seem likely that the origin of Hominidae lies somewhere within Ramapithecinae, the subfamily created to contain these forms. Exactly which ramapithecid is most closely related to hominids is not known, and it is highly unlikely that an actual hominid ancestor is represented among the forms discovered so far.

Although we are still thus in the dark as to the details of hominid origins, we can propose a reasonably satisfying scenario of the rise of the ramapithecines, along lines suggested by David Pilbeam. It seems that in the early Miocene hominoids existed only in Africa, which had not yet made physical contact with Eurasia. At this time Africa was largely covered with humid tropical forests, in which early apes lived off fruits, leaves, and so forth. Early in the middle Miocene, however, perhaps around 15 million years ago, several events occurred that were of major importance for hominoid evolution. Africa/Arabia made contact with Eurasia, allowing the faunas of the two continents to intermingle and shaking up the ecological equilibrium that had been established in each. Hominoids were able to penetrate Eurasia, and middle Miocene hominoids have been found in a broad area stretching from southern Europe to southwestern China. At the same time a global cooling seems to have set in, and the ramapithecines apparently occupied a habitat which was rather different from that of their African precursors of the early Miocene. Their fossils have been found in associations suggesting tropical to temperate woodland rather than continuous dense forest.

Peter Andrews has pointed out that both *Sivapithecus* and *Ramapithecus* were large-bodied by primate standards and that the discontinuity of the woodland habitat means that such large primates cannot have been exclusively, or perhaps even mainly, arboreal. Further, he notes that woodland environments are more seasonal in fruit production than is tropical forest. Putting the two observations together, he suggests that the ramapithecines must have been omnivorous and at least partly terrestrial. The adaptations of large-molars and thick dental enamel in the ramapithecines plausibly reflect this environmental and behavioral shift, since together with the canine reduction that some of these primates also show, these features can—if one wishes—be interpreted as a functional complex adapted for the processing of tough vegetable foods of the kind more typical of a terrestrial or partly terrestrial diet. The ancestral great apes, on the other hand, retained the thin-enameled, relatively small-molared condition—adapted for chewing softer plant foods—appropriate to their tropical forest habitat.

Ramapithecines are known over a span of many millions of years. Indeed, one genus, *Gigantopithecus,* survived in China until well under a million years ago. However, the period between about 8 million and 4 million years ago is very poorly represented by fossils, and we know virtually nothing about the doubtless complex set of events in hominoid evolution that this period must have witnessed. By the time we get to Hadar, hominids were up and walking around, and had been for who knows how long.

The only hominid known from the period between 3 million and 4 million years ago is *Australopithecus afarensis,* the concept of which, as we have seen, has not gone unchallenged. The fossil assemblage consists of jaws, teeth, postcranial bones, and skull fragments from Hadar at time levels ranging from about 4 million to 3 million years and of the jaws and teeth from Laetoli, about 3.7 million years old. Johanson and White believe that all of this material belongs to the same single hominid species, which at present is unknown from anywhere else except, perhaps, the lower levels at Omo. Others believe that more than one species is represented in the assemblage, or that the one species present is already known from South Africa in the form of *Australopithecus africanus.*

One couldn't ask for a starker illustration than this of the problems involved in recognizing fossil species. And as we have seen, there can be no totally objective answer to the question How many species? In one way this instance is unusual, however: the question here involves size as much as it does shape. The most striking variable among all the fossils assigned to *afarensis* is size. The size difference between the biggest and smallest individuals in the sample is very considerable by almost any standard and might be explained in any of a number of ways. Two basic possibilities stand out, however. Either we are faced with an assemblage in which the larger individuals represent one species and the smaller ones another, or we have evidence of a single highly variable hominid species. In the latter case, the simplest explanation for such variability would be sexual dimorphism, the larger individuals being males and the smaller ones females; presumably the size ranges of the two sexes approached each other or even overlapped in the middle. If dimorphism is indeed the explanation, its degree would be remarkable but not completely out of line.

One of the interesting things about the morphological variation of the Hadar/Laetoli fossil hominids is that it appears not to be consistent with size. One or two differences in morphology between large and small specimens do seem to be due to "allometry," as the factor underlying size-related shape differences is called. For example, the tiny Lucy has a rather V-shaped lower jaw as a result of its narrow width in front. But this is said to scale down arithmetically from specimens of large and intermediate size. Similarly, the shaft diameter of the larger long bones is relatively greater than that of smaller ones, since strength increases only with the cross-section of the bone and the weight borne increases with body volume. The latter especially is a typical allometric difference, and many distinctions between males and females of species with high body-size dimorphism are normally attributable to the same cause. But in *afarensis* some variations, for example in the form of certain teeth, seem unrelated to size, while there is no necessary variation in many characters between individuals of very disparate sizes. The upshot is that the size dimorphism of *afarensis* appears to carry along with it rather little obligate shape differential. It is possible that this hints at the presence of more than one size-variable species in the sample, but for the moment it seems reasonable to accept Johanson and White's conclusion.

From our point of view, however, perhaps the most significant aspect of variation in the *afarensis* sample is that it is not consistent with time. Insofar as the comparability of parts allows us to judge, differences between individuals sampled at the same time level seem to be at least as great as those between individuals sampled early and late in the time series.

We believe that Johanson and White are probably correct in assigning all of the Laetoli and Hadar hominids to the same species. Accepting this, can we sustain the claim that this species is distinct from any other previously described? It has been argued that the features quoted by Johanson and White as distinguishing *Australopithecus afarensis* from *Australopithecus africanus* do not in fact consistently differentiate between the two, and that the Laetoli and Hadar samples represent no more than separate subspecies of *Australopithecus africanus*. On the other hand, it has also been argued that at least some of the Hadar material is not only unlike *Australopithecus africanus* but in fact shows affinities with *Paranthropus*. The several hundred Ethiopian fossils involved have not yet been comprehensively described, and it is clear that consensus is some time away. Our own judgment, however, is that separation of *afarensis* from *africanus* is indeed justified, whether or not Johanson and White's diagnosis is adequate to show it. Further, we believe that the allocation of the species *afarensis* to *Australopithecus* (as opposed to *Paranthropus* or *Homo*) is also appropriate, at least if the generic designation is to be derived from names already current. Whether or not *Australopithecus afarensis* subsequently gave rise both to an *Australopithecus africanus* → *A. robustus* lineage on the one hand and to a *Homo habilis* → *H. erectus* → *H. sapiens* lineage on the other is an entirely separate question.

In the period between about 4 million and 3 million years ago, then, we have substantial evidence for at least one hominid species, *Australopithecus afarensis*. *Australopithecus afarensis* was highly dimorphic sexually; the smallest females were not much over three feet high and perhaps fifty pounds in weight, while large males may have been up to about four and a half feet tall and weighed perhaps a bit over a hundred pounds. Relative brain size was probably a bit less than in *Australopithecus africanus*, and although *afarensis* was without question an efficient upright walker there is absolutely no evidence that this hominid made tools. In fact, the presumption has to be that it did not. Nonetheless, *afarensis* was undoubtedly a highly successful primate; the species persisted apparently unaltered for a span of time approaching a million years. This is our first clue in hominid evolution as to *pattern:* no matter whose interpretation we accept of the actual relationships involved, the first million years of documented hominid evolution (fully a quarter of the adequate record) is marked not by change but by its absence.

The period between 3 million and 2 million years ago is a little less well sampled than the million years preceding, but the fossils are almost inevitably farther-flung. Primarily, we have from this time the hominids from Makapansgat, Sterkfontein (main level), and Taung, in southern Africa, and the hominids from Omo compared by Clark Howell to *Australopithecus africanus* from the South African sites (although the earliest Omo materials, about 3.1 million to 2.9 million years old, may be *afarensis*). The Omo fossils consist of various isolated teeth and fragmentary jaws and bones dating between about 3 million and 2.5 million years, and possibly as young as 2 million.

This period—3 million to 2 million years ago—is best represented, then, by classical *Australopithecus africanus* from South Africa. As in the case of *Australopithecus afarensis*, there is no evidence to support the making of tools by *Australopithecus africanus*, although if the hominids were assembled at the sites by carnivores there would be no reason to expect the fossils to be accompanied by artifacts. It has been estimated, given the eruption and wear of their teeth, that the hominids found at both Sterkfontein (about forty individuals) and Makapansgat (some seven to twelve individuals) averaged about twenty-two years of age at death, and it has been remarked that there are few infants or old individuals represented in the samples. This is perhaps surprising if the bones were accumulated by carnivores, since it is easier for predators to prey on the more defenseless members of a group of any sort of mammal than on those in the prime of life. Whether this throws doubt on the origin of the accumulations and therefore enhances the significance of the lack of tools is not clear, but what does seem to be well established is the absence of tools at Omo before about 2 million years ago.

What we know otherwise of South African *Australopithecus africanus* is similar to what we know of *afarensis*: open-country environment, bipedalism, average brain size under 500 cc. (but small body size), and so forth. Whether sexual dimorphism was as marked in *africanus* as in *afarensis* seems unlikely, although some of the jaws from Makapansgat are pretty big. The braincase of these hominids was relatively delicate and rounded, the browridges smallish, but the face rather projecting, especially below the nose. It is possible, however, that this last feature may have become a little exaggerated through being based too closely on the best preserved *africanus* cranium, that of a presumed female from Sterkfontein. This specimen has a surprisingly "dished" profile, which may have been produced to some extent by postmortem distortion. The front teeth of *africanus* were relatively large compared with the back ones, which were themselves large compared with body size. Behaviorally, a good bet is that these hominids were nomadic scavengers and gatherers, living in small, mobile groups that ranged over large areas. Whether weaponless bipeds would have—or could have—hunted is a moot point; hominid bipedalism seems on the face of it ill-adapted for hunting without projectiles (although it did in the end turn out to be advantageous for hunting with them), and it is becoming increasingly clear that bipedalism must have evolved in quite another context. Of course, we shall never know whether or (more probably) for how long hominids had used tools made from perishable materials such as wood before they began to shape stones.

If the morphology of *africanus* is reasonably well known, however, its longevity remains something of a question. And the question brings us back again to species recognition. It seems reasonably well established that the South African material may cover a period of up to half a million years; certainly the Omo fossils do. The Omo hominids attributed to *africanus* may, indeed, span a period twice that long, and if, as some scholars suggest, some early Olduvai bones are assignable to *africanus*, the species was around throughout the million-year period we are discussing. However, much of this is rather slender evidence, and for the moment it seems safest to say that *africanus* seems to have been around for a pretty long time, rather than to try to put a specific minimum figure on it.

The most striking innovation of this period between 2 million and 3 million years ago, at least that we have direct evidence for, is the introduction of stone tools. As we have seen, these turn up for the first time at Hadar, unfortunately at a level of the sequence that has yielded as yet no hominids. These tools, some half-million years older than the previous oldest known, from Omo, are not yet known in quantity. However, they seem to be broadly comparable to the Oldowan of early Olduvai, with "chopper" tools consisting of shaped cores, and small sharp flakes which may have been used for slicing. The Hadar tools occur long before *Homo* is known to be present in Africa, and pose an obvious question: who made them? Are they the products of *Australopithecus,* or do they suggest that *Homo* was around in the Afar as early as 2.5 million years ago? Since, as we have said, the *known* fossil record provides only *minimum* dates for the appearance of fossil genera or species, the question remains open. But if the first alternative is the correct one, as it may well be, either it suggests the existence of a further, as yet unknown, species of *Australopithecus*, one perhaps antecedent to *Homo*, or it points to a phenomenon that we find emphasized elsewhere in the paleoanthropological record: the decoupling of technological advance from morphological innovation in human evolution.

From this point on, dividing hominid evolutionary history into million-year periods becomes rather less useful as the picture becomes more cloudy. Not the least of the troublesome aspects of crossing the 2-million-year mark is that it brings us face to face with the earliest forms that have been attributed to our own genus, *Homo*. The question of what should or should not be included in *Homo* is a vexed and inevitably emotive one. Louis Leakey's *Homo habilis* was, as we have seen, based on less than entirely satisfactory evidence and owed its initial existence essentially to Leakey's preconception that very early *Homo* had existed in Africa. Nevertheless, with the discovery of the relatively large-brained ER-1470 skull, which appeared not to be *Australopithecus* but at the same time was clearly not *Homo erectus*, the reality of *Homo habilis* seemed to be borne out. To be iconoclastic for a moment, however, we should point out that in fact even *Homo erectus* is substantially more different from *Homo sapiens* (the first-named [type] species of the genus and therefore the one with which all other potential members of the genus are to be compared) than are most if not all other mammalian species from the type species of their genera. Of course, we do not wish to annoy our colleagues by suggesting the renaming of any fossil hominids, but we do think that it is wise to bear in mind that the standards of generic classification normally applied in mammalian studies are not necessarily those applied these days in paleoanthropology. It seems almost as if to exclude various early hominids from *Homo* smacks of discrimination to the good-hearted liberal sentiments of many paleoanthropologists. We will run into the consequences of such feelings again, later in this story.

In any event, virtually all nonrobust hominid fossils less than about 2 million years old are widely if not universally regarded today as belonging to the genus *Homo*, and the earliest of this material is assigned by most authors to the species *Homo habilis*. ER-1470, about 1.9 million years old, has become the archetype of *Homo habilis* in East Africa, and with its apparently flattish face and rather well-inflated cranium that contained a brain of about 775 cc., this specimen does not at first glance possess the gestalt of *Australopithecus africanus*, the species to which it might alternatively be assigned. When other East Turkana specimens are introduced, however, this overall picture changes somewhat. One of the two skulls found in 1973 was a quite complete cranium of a gracile individual with a brain volume of little more than 500 cc., thus only very slightly above the range of those recorded for the Sourth African *Australopithecus africanus*. The geological horizon from which the fossil was recovered is somewhat more recent than that which yielded ER-1470, dating to perhaps 1.7 million years ago. In all respects except for the modest projection of its face this skull, ER-1813, closely resembles the best-preserved *africanus* skull from Sterkfontein—and as we have noted, this latter specimen may be a little distorted in the facial region. Yet ER-1813 is also sufficiently similar to ER-1470 to have been considered by some scholars as a female *Homo habilis*, of which they view ER-1470 as a male. Does this mean that all of this material should be assigned to *Australopithecus africanus*? Or that *Homo habilis* was sexually quite strongly dimorphic in brain and body size? Or that *Homo habilis* and *Australopithecus africanus*, two hominid species whose skulls were distinguished largely if not entirely by the size of the cranial vault, lived along the shores of the ancient Lake Turkana at about the same time?

The evidence from East Turkana is not conclusive on this point. Ultimately, analysis of the postcranial fossils may help decide, although their evidence seems rather equivocal. A couple of thighbones from the lower part of the Koobi Fora Formation, approximately contemporaneous with ER-1470, have large heads and short necks, like modern man, while a couple from higher in the section and perhaps about the same age as

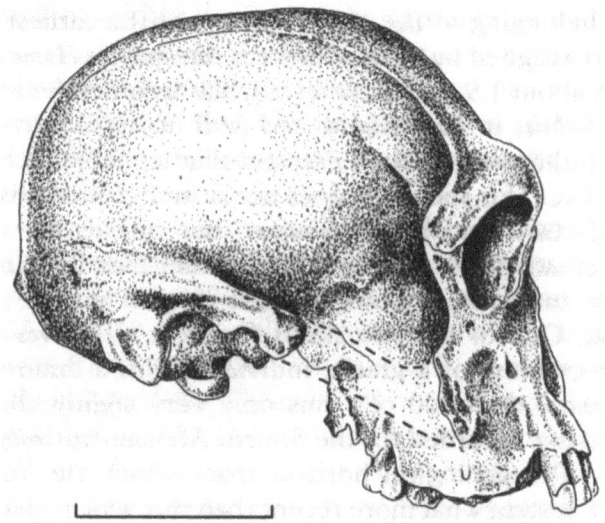

The ER-1813 cranium from East Turkana. The scale represents five centimeters.

ER-1813, seem to have small heads and long necks, like *Australopithecus*. But this is insufficient evidence upon which to opt for the third of the alternatives given above, especially since more than one species of *Australopithecus* was present at East Turkana, and the analysis of the material is anyway incomplete. One must, then, look to other parts of the world for evidence to support or deny the validity of *Homo habilis*.

Starting only a few hundred miles away, we have, of course, Olduvai. The evidence from Bed I and Bed II certainly indicates the presence at Olduvai between about 1.8 million and 1.6 million years ago of at least one hominid apparently similar in comparable parts to *Australopithecus africanus* but with a rather larger brain than any known from the earlier deposits in South Africa. From later South African localities, however, have also come gracile hominids with larger brains than those of classic *Australopithecus africanus*. From Swartkrans, and dated at probably around 1.9 million years come some partial skulls and dentitions that are demonstrably not robust *Australopithecus* but that do appear to possess more voluminous brain cavities than *africanus*. Similarly, a new find from a late part of the Sterkfontein site (probably about 1.7 million) is said to be somewhat different from *africanus* and to have a larger brain.

The Javanese hominids recovered by Dubois and later workers are now known to span a considerable period of time, back to almost 2 million years ago, and perhaps more. Not many Javanese hominid fossils are known from the earliest part of the sequence, and the earliest of them, consisting of most of a skull minus the face, is juvenile and thus hard to interpret. However, its cranial capacity was some 650 cc., and this would of course have increased (although not much) had the individual survived to adulthood. Slightly younger are parts of a skull and some lower jaws; this material was early on compared closely with Olduvai *Homo habilis*, and resemblances are apparent. An estimate of brain size from the preserved back of the skull suggests that it was not much smaller than that of ER-1470.

One of the reasons why most paleoanthropologists have not found much difficulty of late in accepting all or most of this material as *Homo* is that by 2 million years ago stone tools were well established in the archaeological record, and stone tools are regarded by many as the hallmark of our genus. The levels at Sterkfontein and Swartkrans that yielded putative *Homo* have also produced stone tools, as of course have Olduvai, the post-2-million-year levels at Omo (whence has come a small sample of isolated teeth which have been assigned to *Homo habilis*), and East Turkana. However, the problem with using stone tools as markers of this kind is that it is impossible to know who actually made them; and at a time level some half million years after we know that stone tool making was introduced among hominids, and during which we know that more than one hominid was around, it seems unwise to use these (Oldowan type) tools as diagnostic of anything.

So what does the agglomeration of material commonly assigned to *Homo habilis* actually represent? We believe that at present it is impossible to be sure. We know that in the period following about 2 million years ago we find a good deal of variability in brain size among gracile hominids, together, apparently, with the general retention of a rather unspecialized structure of the skull. But whether we are dealing with one species, or two, or many, is difficult to tell. And without a reasonable ability to discriminate between species, we would be unwise to attempt to infer pattern. For the present, then, we would choose to retain *Homo habilis* as a species of convenience, a term useful for identifying a particular suite of fossil material, but to which we would not attach an inordinate amount of biological significance.

At the same time as it presents us with these uncertainties, however, the period of 2 million to 1 million years ago does provide us with the first absolutely unequivocal evidence that we have of the coexistence of at least three separate hominid lineages, and also with a nice example of stasis in the hominid fossil record. For sites all over Africa have yielded evidence not only of gracile hominids but also of robust ones. From Swartkrans and Kromdraai in South Africa has come the modestly robust *Paranthropus robustus,* and from eastern Africa has come

Paranthropus boisei. The Kromdraai sample of *robustus* seems on best estimates to date from around 1.5 million years ago, while the older robusts from Swartkrans seem to span the period of maybe 1.9 million to 1.7 million years. This minimal sampling hardly gives us an adequate idea of how long this species survived in South Africa, although it does provide clear evidence of contemporaneity with the eastern African robusts, and the evidence for long persistence of *boisei* is compelling. Robust hominids appeared at the Omo over 2 million years ago and lasted a considerable time there.

At East Turkana the story is similar: *boisei* appears early on, below the KBS tuff, and continues to the top of the collected section, at about 1.2 million years. At Olduvai, Zinj occurs near the base of Bed I, and isolated teeth and bone fragments are found at levels that continue close to the top of Bed II, indicating a minimum time span of about three-quarters of a million years. Other occurrences in eastern Africa fall within the range of slightly over 2 million to 1 million years. Oddly enough, for what was without doubt a terminal lineage that coexisted (and presumably competed) throughout its time span with possible or undoubted species of *Homo, boisei* fossils are relatively rare in the lower strata at East Turkana (whence the species is best known) but become more common than *Homo* in later horizons. In any event, *boisei* springs forth fully fledged and hyper-robust at its first appearance in the geological record and continues unchanged as far as one can tell for the million years until its last occurrences. Its South African contemporary, *robustus*, is more conservative than *boisei*, and it is even possible that the latter species was derived from an early population of *robustus*; if so, this is a good example of the persistence of an ancestor alongside its descendant. When the two species split must remain conjectural, although this must have occurred earlier than 2 million years ago.

Virtually since its initial discovery *Homo erectus* had been thought of essentially as the hominid species (or stage) of the middle Pleistocene (a period now generally understood by some as beginning about 700,000 years ago but by others as commencing at 1 million)—until 1975, that is, when East Turkana produced a finely preserved cranium of *Homo erectus* in deposits dated to about 1.6 million to 1.5 million years B.P. (before the present). It had for some years been realized that the *Homo erectus* skullcap from the top of Bed II at Olduvai might be over a million years old, but the astounding age of the new East Turkana specimen, ER-3733, put it right at the beginning of the Pleistocene, at about the time when the first global climatic cooling of that epoch was getting under way. Although this cooling episode did not lead to the advance of an ice sheet in Europe (it did, however, in North America), it clearly had a worldwide climatic impact, and the environmental fragmentation which this must have involved presumably provided plenty of opportunity for the isolation of populations on which speciation so largely depends.

In any event, the new specimen, ER-3733, is clearly *Homo erectus* in its most typical form. So classic is it, indeed, that its describers, Richard Leakey and Alan Walker, compared it closely to *Homo erectus* from Choukoutien, a form over a million years younger. The braincase is largish, with a capacity of around 850 cc., the browridges are jutting, and the forehead rises aft of a trough behind the ridges. The cranial vault is long, weakly inflated, and angled at the back, and the bone composing it is thick. The face is relatively reduced, and the teeth, as far as can be deduced from largely empty sockets, were likewise. This description could equally be applied to any other of the well-preserved skulls of *Homo erectus* known from around the world in the period between about 1.6 million and 0.4 million years B.P. We have in the past argued that for *Homo sapiens* to have been derived from *Homo erectus* would have involved acquiring this specialized skull form and then losing it again, since the thin-walled, rounded, and high-vaulted crania of *Australopithecus africanus/Homo habilis* and *Homo sapiens* in several ways resemble each other more closely than either does *Homo erectus*; we have therefore wondered aloud whether Louis Leakey might not have been right in placing *Homo erectus* on a limb and linking the others. By now, however, the evidence is overwhelming that some *Homo erectus* population must have been ancestral at some remove to *Homo sapiens*, for *Homo erectus* seems clearly to be the standard-issue hominid for the million years following 1.5 million B.P. Let's look at the major pieces of evidence.

Shortly after the recovery of ER-3733 a rather similar but slightly less complete hominid skull, ER-3833, turned up at East Turkana. This specimen, about the same age as ER-3733, is a little more robust but is clearly the same kind of creature. The face and browridges are a little more massive, the trough behind the browridges is less well defined, and the angle at the back of the skull is not quite so sharp in profile. But nobody would have any trouble in identifying this cranium as that of a *Homo erectus*, and in general ER-3833 resembles a bit more closely than does ER-3733 the approximately 1.2-million-year-old *Homo erectus* braincase from Olduvai Bed II, the cranial capacity of which is roughly 1,100 cc. A much younger East African cranium, probably about half a million years old, comes from Ndutu, in Tanzania; this is also said to have certain points of specific resemblance to Choukoutien *Homo erectus*, but it may fit better with a later group. A few postcranial bones from Olduvai confirm that African *Homo erectus* had a bodily anatomy similar to ours except in the great robusticity of its bones.

Other African hominid fossils from before the half-million-year mark are limited to some relatively late jaws and teeth which are not very helpful in assessing the status of their former owners. The remainder of the firm evidence of *Homo erectus* is Asian, specifically from Java and China. As we have seen, the earliest material from Java is of very considerable antiquity, and some of it falls into our convenience category of *Homo habilis*. Dating of the later hominids is rather approximate, but material from the Trinil faunal zone (equivalent to the Kabuh Formation, from which most specimens other than Dubois' original material come) seems to fall between about 1.0 million and 0.5 million B.P., or perhaps somewhat earlier. The major cranial features are those already listed for the Kenyan specimens: long, low, thick-walled skulls with projecting browridges, angled occiputs, and so forth. Cranial capacities of the seven braincases now known range from about 800 to about 1,000 cc. (for comparison, cranial capacities of modern man run from under 1,000 to about 2,000 cc.).

The earliest potential *Homo erectus* fossils from China are limited to some incisor teeth dated at about 1.7 million years, and are clearly inadequate for any attribution more precise than Hominidae. However, a cranium from Lantian, with a cranial capacity of about 800 cc., may date back to 700,000 B.P., but is probably younger. The skull is quite massive, and the rather underinflated cranial vault reflects the small (for *Homo erectus*) brain. Thanks to the size of the sample recovered and to the meticulous labors of Franz Weidenreich, the Choukoutien hominids still represent despite their disappearance the best-known *Homo erectus* population known from anywhere and remain the archetypes of the species. They are dated nowadays, without great precision, at around 450,000 B.P., and range in cranial capacity from about 900 to over 1,200 cc.

In sum, during the period that lasted from about 1.6 million to 0.4 or 0.5 million B.P., nonrobust hominids seem to be represented by a single species which is both geographically and locally variable, but which has an instantly recognizable gestalt. The major cranial characteristics of this species, *Homo erectus*, have already been enumerated, and postcranially it is clear that *Homo erectus* was robust but an erect biped in the manner of ourselves. What many have found remarkable is that over this long span of time, well over a million years and perhaps as long as 1.2 million, *Homo erectus* shows virtually no change; local and geographical variations are at least as striking as differences between older and younger members of the lineage. Some scholars have suggested that brain size does show an increase over time, pointing to the fact that ER-3733 had a brain of under 900 cc., while the largest of the late Choukoutien population had a brain volume of over 1,200 cc. One should point out, however, that after the East Turkana specimens the oldest firmly dated *Homo erectus* is the Olduvai skullcap, dated at about 1.2 million years and which has a capacity of almost 1,100 cc., larger than all but two of the Choukoutien specimens, which are the best part of a million years younger. Indeed, a recent attempt to quantify variation in *Homo erectus* over time has failed to show significant trends that would convincingly suggest that the species was undergoing any gradual transformation.

Once again, then, we find that a substantial stretch of human physical evolution was marked not by change but by stasis. *Homo erectus* was evidently a highly successful species which spread not only through the tropical and subtropical parts of the Old World but which also penetrated the temperate zone (as, for example, China), apparently the first hominid to do so. To what extent this occupation of a new climatic zone depended on improved technology is not clear (although China gives us the first evidence in the archaeological record of the use of fire by hominids), but it may be significant that the first appearance in East Africa of *Homo erectus* broadly coincides with the equally abrupt initial occurrence in the region of stone tools belonging to the Acheulean hand-axe tradition. In this technique a stone "core" was prepared, off which a flake was struck which served as the tool. This development was unanticipated in the Oldowan stoneworking tradition which preceded the Acheulean. The "bifaces" characteristic of the Acheulean turn up regularly over a long period in Africa, western Asia, and western Europe but are rare in eastern Europe and eastern Asia, where "chopper" tools remained the norm until quite recent times. Even in Africa the arrival of *Homo erectus* and Acheulean tools did not witness the end of Oldowan-type traditions; ER-3733 and ER-3833 derive from the period when the Karari industry was characteristic at East Turkana, while at Olduvai the Developed Oldowan, to which the Karari seems to have been roughly equivalent, survived for hundreds of millennia side by side with the Acheulean. Quite evidently, there is here no clear-cut correlation of culture with hominid physical type. It does seem to be true, however, that the kinds of raw material available had a distinct effect on the composition of stone tool kits, and this may serve to obscure the issue somewhat.

Among the most egregiously counterproductive myths that paleoanthropologists have chosen to inflict upon themselves is the myth of "morphological space." One of the earliest objections to *Homo habilis*, for example, was that there was "insufficient morphological space" between *Australopithecus africanus* and *Homo erectus* to admit an intermediate species, and people took this sort of argument seriously even though it should have been blindingly obvious that *Australopithecus africanus* and *Homo erectus* were two very different hominids and that one could not possibly have suddenly transformed directly into the other.

A similar sort of reasoning seems to have applied in studies of the gray area into which we move after leaving behind forms that are clearly recognizable as *Homo erectus*. Following about 0.4 million B.P. we have a sampling of hominids from many parts of the world that are clearly not *Homo erectus* but yet do not resemble modern *Homo sapiens* either. And even though a very solid case can be made for saying that if one had to choose between placing them in *Homo erectus* or in *Homo sapiens* the obvious choice for these hominids with their big faces, heavy browridges, and long crania is the former, they are instead conventionally classified as *Homo sapiens*. In deference to the profound differences between these fossil forms and modern men, however, they are referred to as "archaic *Homo sapiens*." This is quite evidently not a very helpful or constructive stance to take in any attempt to understand these hominids, but it does have the advantage of eliminating the necessity of figuring out how many species are represented in the overall fossil assemblage, and of what to call them—no small bonus.

At a deeper level, however, the inclusion of these forms in *Homo sapiens* seems to reflect not only the myth of a restricted morphological space between *Homo erectus* and modern man, into which paleoanthropologists have been reluctant to intrude another species, but also the generous liberalism we spoke of earlier, which makes excluding large-brained hominids from our own species or genus seem somehow nastily discriminatory.

Perhaps a further reason for the extension of the concept of *Homo sapiens* well beyond any real biological utility may be sought in yet another anthropological myth: that of the inordinate variability of modern man. As human beings we are naturally finely attuned to the geographical variations evident in our own species, and we tend to think of them as being remarkable. But as far as external characteristics are concerned there are other "polytypic" primate species that show at least as much variation as we do. And, more important, from the point of view of comparisons with the fossil record, modern human bone structure is in fact relatively uniform, as the eminent anthropologist W. W. Howells has pointed out. All modern people have high, well-inflated crania, with parallel sides and rounded occiputs. Browridges are small or most commonly nonexistent, and the forehead rises steeply above the eye sockets. The face is small, and is tucked in below the front of the cranial vault. Perhaps above all, the modern skull is thin-boned and delicately constructed. Of course, if we look at the human cranium in sufficiently fine detail we can find differences in it between geographical groupings of humankind. One group tends to bulge more above the eyes; another shows broader

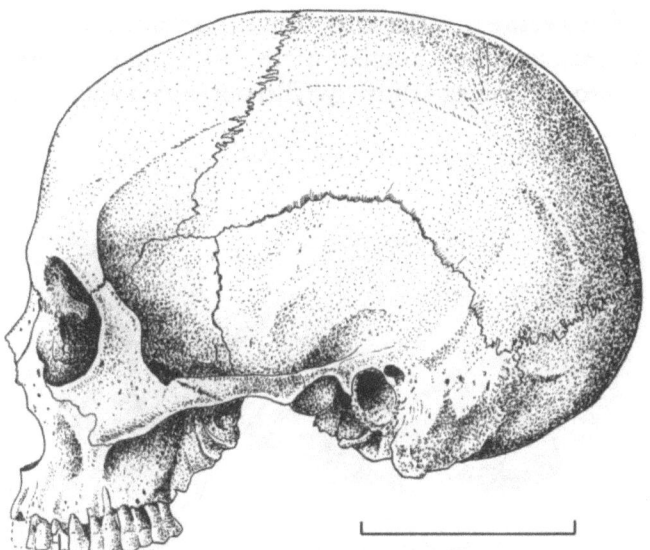

The cranium of a modern man from Thebes, Egypt. The scale represents five centimeters.

cheekbones, hence flatter faces. But these are minor differences, and only serve to underline the homogeneity of modern man relative to what we find among later Pleistocene hominid fossils.

The period we are considering now witnessed the spread of hominids into regions of rigorous climate, and for the first time the concentration of paleoanthropologists and archaeologists in Europe begins to be reflected in the number of fossil hominids found there. Specimens such as Schoetensack's Mauer jaw have traditionally been regarded as *Homo erectus*, but recent studies indicate that this species is not convincingly represented in Europe and that most likely all European fossil hominids belong instead to a post-*erectus* phase of human evolution. The Mauer specimen is probably the earliest of these fossils but as an isolated lower jaw must remain problematical. Probably a little younger is a cranium from Petralona, in Greece, whose dating is highly uncertain but estimates of which center around 350,000 B.P. This skull, with a brain volume of about 1,200 cc. (only a little below the modern average), has a large face with massive browridges inflated by capacious sinuses. The cranial vault is long and low, angled at the back, and tremendously thick-walled. A recent study by Chris Stringer allies the Petra-

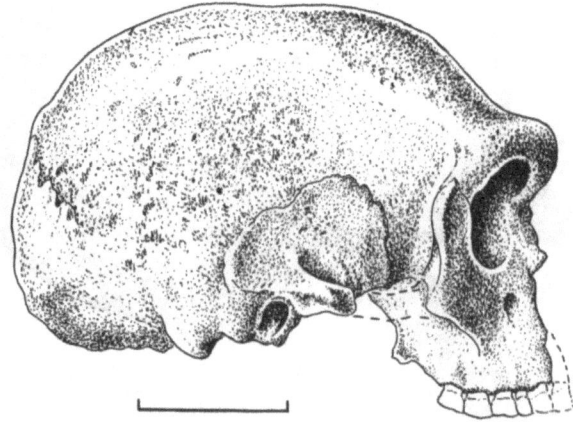

The Steinheim cranium from Germany. The scale represents five centimeters.

lona specimen with a suite of other material from Europe, most of which is probably a little younger. Dates range from the late Mindel glacial to the succeeding interglacial or the early Riss (around 0.4 million to 0.2 million B.P.). This material includes a face and some lower jaws from Arago, France, an occipital bone from Verteszöllös, Hungary, parts of a cranium from Bilzingsleben, East Germany, the back of a skull from Swanscombe, England, and, the only more or less comparable specimen in terms of its completeness, a cranium from Steinheim, West Germany. The Steinheim skull, with a cranial capacity equivalent to the Petralona, resembles the latter particularly in having a massive face, but the browridges are slightly smaller and the cranial vault is somewhat rounder. The Arago face is quite a good match for the Petralona specimen, however.

Perhaps surprisingly, the best matches for the Petralona skull come not from Europe, but from Africa. A fossil face found not long ago at Bodo D'Ar, in Ethiopia, shows a very similar development of the browridges, and the same great breadth and massiveness of the face, together with other points of sim-

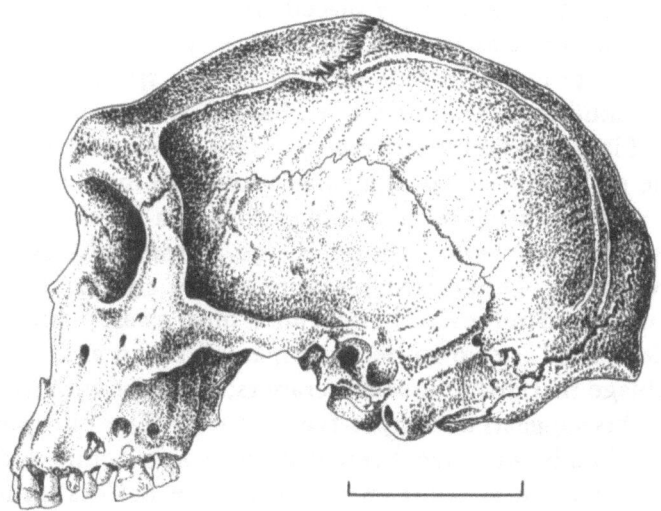

The Kabwe cranium from Zambia. The scale represents five centimeters.

ilarity. This is also true of the cranium (Rhodesian man) found at Kabwe, Zambia, in 1921, which has a brain volume of a little under 1,300 cc. but somewhat thinner cranial bones than those of Petralona or Bodo. A skullcap from Saldanha, South Africa, resembles the Kabwe skull and may be around the same age, although its dating is highly insecure. The age of the Bodo specimen can only be estimated with great imprecision as lying somewhere in the middle Pleistocene; the Kabwe skull is now believed to date from at least 130,000 B.P. and is probably older still. A skull with a damaged face from late deposits at Laetoli estimated to be about 120,000 years old, has recently been described. Its cranial capacity is on the order of 1,200 cc., and although the specimen has been described as quite modern in morphology (meaning, presumably, unlike *Homo erectus*), it has a long, low, thickish vault, a broad face, and pronounced browridges well defined above each eye like those of the other crania we have been discussing.

Materials from Asia of around this age are less abundant yet. A series of faceless crania from the Ngandong faunal zone of Java, once thought to belong to the "archaic *Homo sapiens*" group, has now been firmly placed with *Homo erectus*, and may indeed date back as far as a half million years; if it doesn't, it apparently points to a late survival of this species in eastern Asia. A skull from Dali, in China, has recently been described, and is of late middle Pleistocene age. This specimen has a cranial volume of 1,120 cc., a long cranial vault with moderately thick bone, a broad face, and browridges that, if a bit smaller, seem to resemble somewhat those of the Petralona/Arago/Bodo/Kabwe group. Other features appear less comparable, however, and the affinities of the specimen must await further study. The Dali skull is said to be accompanied by an archaeological assemblage that includes small scrapers; together with its provenance, this suggests a chopper-type industry. The European and African fossils we have been discussing are found either in Acheulean cultural contexts or with more or less equivalent flake-based industries.

It is only with the late Pleistocene of Europe that we come

to the Neanderthals, the one group of "archaic *Homo sapiens*" that it has become customary (although not universal) to exclude from the direct ancestry of "modern *Homo sapiens*." The Neanderthals comprise a group of fossil hominids of highly characteristic gestalt; all come from sites in Europe, the Near East, and central-western Asia that date from the last interglacial or a bit earlier (perhaps 160,000 years ago) to around 34,000 years ago. These hominids had brains at least as large as ours, housed in an inflated but still long, low, and thickish vault that overlay a broad and rather flat cranial base. The Neanderthal braincase is adorned with strong browridges in front, and is markedly protuberant at the back. The cheekbones sweep backward, and the massive face is extremely prominent in the centerline. The postcranial skeleton, like that of earlier hominids, is extremely robust. It is with the Neanderthals that we first find evidence in the archaeological record of some degree of spiritual awareness, if burial of the dead with flowers (represented now by fossil pollen) or other grave-objects such as animal horns may be so interpreted. Neanderthal fossils are normally accompanied by variants of the "Mousterian" stoneworking tradition, in which preformed flakes were struck from a core with a hammerstone, then trimmed to produce a variety of scrapers, knives, spear points, and so forth. Like *Homo erectus* and the other "archaics," the Neanderthals were accomplished big-game hunters (although it may be unwise to exaggerate the significance of meat in their diet), and they presumably lived in small, mobile bands that covered quite large ranges and may, indeed, have followed the movements of large migratory mammals.

But as successful as the Neanderthals appear to have been in the earlier part of the last glaciation, they disappeared abruptly in Europe around 35,000 years ago, to be replaced by men of fully modern aspect. As far as Europe is concerned, the pattern seems clear: sudden displacement (and presumably extinction) of a resident form by an invader. This is a kind of episode that has been repeated over and over again during the history of life on earth, but many paleoanthropologists have been reluctant to accept it as the explanation for the demise of the Neanderthals. Some have proposed that the European Neanderthals evolved directly into man of modern aspect; others have suggested more cautiously that the invading moderns interbred with the Neanderthals and absorbed their physical characteristics. But in Europe, at least, it is inconceivable that either occurred. In the first case, the pattern of abrupt displacement is too regular and too short-term to allow in situ evolution by any mechanism whatever; and there are no European intermediates. In the second, the morphology of the invading moderns makes no concession at all to that of the Neanderthals; and even in the (to us highly improbable) event of full genetic compatibility between Neanderthals and moderns, it is inconceivable that Neanderthal characteristics would have been so completely masked out.

Elsewhere than in Europe, things are less clear-cut. Certain northern African specimens, such as a skull from Djebel Irhoud, in Morocco, show some affinities to the European Neanderthals, to which group they might for convenience be assigned. Earlier than such specimens are two partial skulls from the Kibish Formation, in the Omo; these may be as old as 130,000 years but could be a good deal younger. In any event, although the two specimens were recovered from more or less the same geological level, they differ quite strongly. One of them has a longish and not very high cranial vault with thick bone and a cranial capacity of about 1,400 cc.; the frontal region is missing, but the back of the skull is somewhat angled. The other is more modern in aspect, with a higher and more lightly built vault with a more rounded profile at the back. It was at first concluded that this constituted evidence of exceptional morphological variability among "early *Homo sapiens*," but the betting is now that two different populations are sampled here. From farther south, at South Africa's Border Cave, comes a fossil sample of several fragmentary individuals that may date back to around 90,000 B.P. or more. These hominids are associated with a rather advanced tool kit, however, and may be only half as old as claimed. A partial adult cranium is said to show affinities to modern

Bushmen, although it does have some modest browridge development.

Fossils of this time period are poorly represented in eastern Asia, although some fossils from fairly late times but of rather "archaic" aspect have been found in Australia, while a modern-looking skull was found in a cave deposit in Borneo that may date as far back as 40,000 B.P. The western part of the continent, however, has produced some suggestive fossils. Two sites in Israel have yielded specimens that appear to show some intermediacy between more "archaic" types on the one hand and modern *Homo sapiens* on the other. A skull from the cave of Skhūl, for instance, has a heavy, projecting face with distinct browridges, if not huge ones. This complex is tacked onto the front of a rather modern cranial vault and the specimen dates from about 35,000 B.P. At the neighboring site of Tabūn, only some 5,000 years earlier in time, the skeletal population is

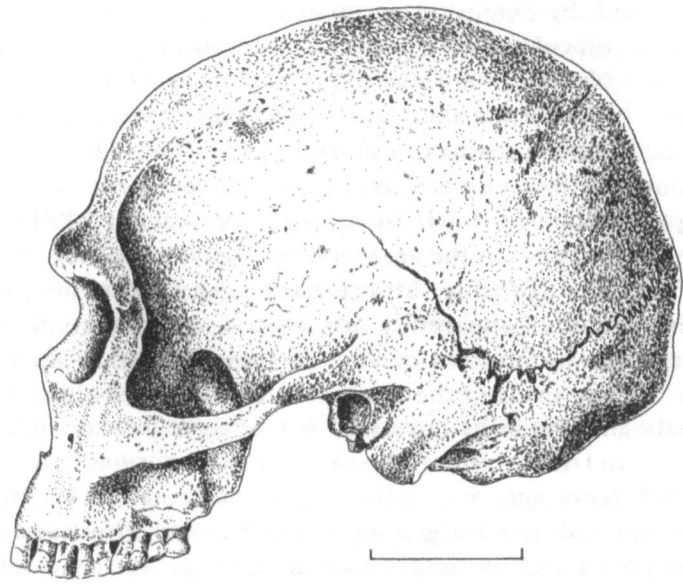

The best preserved of the crania from Skhūl, Israel. The scale represents five centimeters.

pretty Neanderthal-like. A skull from Djebel Qafzeh may be a bit earlier, dating from perhaps 50,000 B.P., but like the Skhūl specimen, it shows the association of a rather rounded vault with a heavy face.

The study of archaeological remains unfortunately does little to clarify the picture. The Skhūl and Qafzeh fossils are associated with the Mousterian tools typical of the Neanderthals, and although it was until recently the case that no Neanderthal had ever been found in association with the sophisticated Upper Paleolithic tools characteristic in Europe of anatomically modern man, such an association is now known from St. Césaire, France, and less certainly from Vindija, Croatia. The Upper Paleolithic toolmaking industries are completely different in character from the Mousterian; typically, long, narrow flakes were struck off carefully prepared cores by the use of intermediate bone or antler punches. Some archaeologists believe that this new tool kit reflects a quantum leap in hunting technique—and, by extension, in mental abilities. Certainly its appearance coincides with a remarkable human population explosion after Neanderthal times. St. Césaire, which boasts both a "classic" Neanderthal skull and a "Chatelperronian" Upper Paleolithic industry, seems to date from around 34,000 B.P., which is about the time modern man came on the scene in western Europe, and before which (or at least, before about 38,000 B.P.) stone technologies in the area were probably uniformly of Middle Paleolithic (broadly, Mousterian) aspect. Whether the St. Césaire skeleton, apparently not a burial, represents the Chatelperronian toolmakers or one of their victims is anybody's guess. In any event, other evidence also indicates that Neanderthals and moderns coexisted for at least a brief period in Europe. In Africa the Acheulean, which survived to about 150,000 years ago, was joined some time earlier by Middle Stone Age cultures roughly equivalent to the Mousterian; but the Border Cave specimens (whose date, as we have seen, is somewhat in doubt) were accompanied by artifacts of Upper Paleolithic type.

If all this (and there's plenty more) sounds confusing to you, take heart. It does to us, too. The last glaciation obviously witnessed a complex series of events in hominid evolution that we are viewing through a very dark glass indeed. During the last hundred thousand years or so we have fossil evidence for highly specialized hominid populations, such as the Neanderthals, that cannot have borne any direct ancestral relationship to ourselves. But in the period before about 35,000 B.P. we also have fossils that in some characteristics approach anatomically modern man while retaining a substantially to predominantly archaic aspect. Finally, we have the appearance, apparently pretty abruptly everywhere, of men of completely modern aspect. Some scholars have perceived evidence for regional transitions from known fossil forms to the modern human populations that occupy the same geographic areas today, but none of these proposed transitions has been convincingly documented. One general observation does stand out, however, as of fairly universal validity: as David Pilbeam has noted, the arrival of fully modern populations everywhere involved a remarkable loss of robusticity in the cranium and in the postcranial skeleton. And this final loss of robusticity seems to have been a rather late event in all parts of the world.

The climatic fluctuations of the late Pleistocene must have provided ideal conditions for the fragmentation and isolation of hominid populations worldwide, with subsequent local differentiation between populations that may or may not have been accompanied by speciation. Such differentiation seems to be reflected in what we see in the fossil record; and unfortunately we cannot tell to what extent this might have obscured a more basic underlying pattern. With the spotty evidence at our disposal we can construct an almost unlimited number of scenarios to account for the final arrival on earth of modern man, and at this point we are unable to make clear choices between any of them. The general pattern, if you will, *is* chaos. Amid all this uncertainty, the "side-branch" Neanderthals provide a rather refreshingly cohesive picture. As W. W. Howells has observed, this distinctive European and Near Eastern population seems to have been well established by the last interglacial period and to have survived in essentially unaltered form until cut off in its prime about 35,000 years ago. This is not by any means the longevity of, say, *Homo erectus*; but in the ferment of the late Pleistocene, 140,000 years or so isn't bad.

Retrospect

What, then, does the hominid fossil record tell us about pattern in human evolution? Does it tend to confirm the idea of gradual directional change, or does it suggest a more episodic sequence of events? Where a clear pattern does emerge, it supports the second alternative, but there are stretches of time when no clear pattern is discernible at all. Lack of pattern is predicted neither by the gradual change notion nor by that of stasis interspersed by rapid change; but it may be fair to suggest that, given fairly regular sampling, pattern is hardest to distinguish when change is at its most rapid. Let's briefly summarize the evidence.

In the period between about 4 million and 3 million years ago we have an assemblage of fossils, from the Afar and Tanzania, that are most reasonably interpreted as belonging to the single species *Australopithecus afarensis*. This small, bipedal, nontoolmaking hominid was relatively small-brained, and as far as can be told persisted unchanged throughout a span of almost a million years. Nontoolmaking or not, this species was obviously well adapted to its open-country milieu, which makes one wonder about the sort of life-style that first brought hominids or proto-hominids onto the open savanna. The next million years of hominid evolution are more spottily sampled, but the successor to *afarensis* seems to have been not greatly different in morphology and adaptation and to have persisted for a similar span of time. This period did, however, see the beginning of the archaeological record, with the first stone tools turning up at about 2.5 million years ago. No hominids are known from the beds in which the tools were found, but there is no evidence among contemporaneous hominids known from elsewhere that this innovation coincided with any morphological change or with the arrival of a new hominid species.

When we pass the 2-million-year mark the situation becomes more obscure, largely because of the problems of recognizing species in the hominid assemblage of around this time. *Homo habilis* is not easy to diagnose, even on criteria of brain size, and we do not believe that it would be helpful to try to generalize about pattern until this material is better understood. In the case of robust australopithecines, on the other hand, pattern is quite evident. *Australopithecus robustus* is not well sampled in time, but its close relative and contemporary *A. boisei* abruptly appeared, fully fledged, around 2 million years ago and persisted apparently unaltered until disappearing equally abruptly about a million years ago. Meanwhile, fossils clearly recognizable as *Homo erectus* turn up, unanticipated, in the eastern African record at about 1.6 million years ago. It is possible that some minor increase in body and therefore also brain size occurred on average in *Homo erectus* over the subsequent million years or more, but it has proven impossible to demonstrate this convincingly, and we find it significant that the earliest *Homo erectus* was most closely compared by its describers to Peking *Homo erectus*, which is not only the traditional yardstick for the species but, at a million years or more younger, is one of its most recent occurrences. Once again, we find here that the appearance of a new species does not coincide with the introduction of a new stoneworking technology, although the Acheulean in Africa turns up not long after *Homo erectus*. Equally striking is that technological advance did not make obsolete the old technology; the Developed Oldowan continued alongside the Acheulean in Africa for half a million years or more. In other parts of the world equivalent unsophisticated industries continued for very much longer.

Following the long tenure of *Homo erectus* the picture once more loses its clarity, partly because dating is difficult and partly because we begin to pick up more morphological variability in the fossil record. If, as has been suggested, the Petralona and Steinheim skull types represent male and female, respectively, of the same hominid, this variability, in the period between about 400,000 and 100,000 years ago, need not worry us, and we have evidence for one single dimorphic type that persisted throughout this period. We feel, however, that the dimorphism explanation is probably oversimplified and that the actual picture is more complex.

In the succeeding 100,000 years events were without doubt more complex, but we are unable to say exactly what was going on. Two sets of data do appear to show a certain amount of regularity in patterning, however. First there is the persistence in Europe, for well over 100,000 years without apparent change, of the distinctive Neanderthal type. Second, there is the apparently rather abrupt appearance of fully modern man. The nature of this event has been obscured somewhat by the inclusion of a hodgepodge of fossils of remarkable variety in the species *Homo sapiens*, and by the uncertainty of dating of much of this material. It does seem, however, that if the arrival of anatomically modern man on earth is defined in terms of the final loss of robusticity, we are indeed looking at a short-term phenomenon, even if it would be unwise to place any numbers on it.

In this context it is interesting to note the views of the neuroscientist George Sacher, who has proposed that many of the properties we view today as being uniquely human—language, for example—appeared very late, and only after the brain had reached its present volume. To Sacher, complex hominid patterns such as social organization, hunting, toolmaking, and so forth, were probably based initially on "inefficient" neural processes that find their equivalents in other mammal species. Sacher suggests that language was in evolutionary terms an instantaneous invention, based on the attainment by the brain of a critical size. The other distinctive behavioral attributes of man then flowed from the vastly improved information processing that this quantum neural leap allowed. Interestingly, this notion ties in rather well with the anatomical changes to the skull that permit modern articulate speech. These involve flexion of the cranial base to permit the development of a pharynx. Modern man shows such flexion while classic Neanderthals do not, and the final rounding of the modern human skull may have been tied in with the bending downward of its base. It seems reasonable to suppose also that hominids, such as that found at Skhūl, with primitively robust faces but rounded braincases, reflect the acquisition by Mousterian populations of language; the actual geometry of events here remains cloudy, however.

In any event, Sacher's scenario, while impossible to prove or to disprove, is an attractive one, and helps explain, for example, how some of the very greatest artistic achievements of mankind—the cave paintings of Lascaux and Altamira, for instance—were made so soon after the appearance of securely dated modern man in Europe—and came almost, it seems, out of nowhere. It also fits well with the conclusions of Richard Klein, whose archaeological studies suggest to him that the modern man who succeeded the Neanderthals about 35,000 years ago were not only vastly better hunters but were their intellectual superiors in every way.

One final word about pattern. By looking in detail at the hominid fossil record, we have seen that the idea of slow, gradual progression is not borne out. But nonetheless, by looking back over the entire span of hominid existence, it is possible to pick out a couple of long-term trends, particularly in the increase in brain and body size. These have often been taken as prima facie evidence of gradual change; but in fact to us they say something else entirely. We have already noted that analysis of well-documented trends almost invariably reveals that they are *between* species, which themselves show little change over long periods. Maybe the existence of these trends in hominid evolution reflects competition between coexisting hominid species: a possibility to which our devotion to gradualist ideas seems to have blinded us.

CHAPTER SEVEN
Patterns in History

THE HUMAN past graduates from "prehistory" to "history" with the advent of writing. To some degree the work of the prehistorian and the historian of early times seems much the same: both are "archaeologists." The prehistorian excavates cave deposits, campsites, hunting localities, and the like, while historical archaeologists exhume the remains of villages and cities. But the artwork and texts available to historians provide immensely richer views of what were rather complicated social systems. So, tempting as it is, it is no easy matter simply to extend our look at patterns of change in prehistory to see if similar patterns are to be found in the historical record.

Yet we do feel that human history over the past 6,000 years or so reveals a continuation of the same basic pattern of rapid innovation followed by far longer periods of little or no change. The prevailing view among historians, however, seems to us to remain an expectation of slow, steady, gradual change—in a word, progress. Historians, as we shall see, have suffered from the same sort of Victorian myths that have afflicted evolutionary biologists.

Consider the words of sociologist Kenneth Bock, in his recent (1980) *Human Nature and History: A Response to Sociobiology*:

> At a more general level, the picture of human experience presented by both social evolutionists and sociobiologists is the very antithesis of what historical records show. In place of a continuous process of sociocultural change, the records clearly indicate long periods of relative inactivity among peoples, punctuated by occasional spurts of action. Rather than slow and

gradual change, significant alterations in peoples' experience have appeared suddenly, moved swiftly, and stopped abruptly. What we call civilization has been a rare and sporadic phenomenon in both space and time. Instead of the uniform or unilinear process of change that might be expected to result from a unity of genetic control within the species, where change has occurred it has obviously taken those manifold forms and produced those varied consequences that constitute the array of real cultural differences that history and ethnography reveal to us. (p. 195)

Bock then goes on to argue even more strenuously against the idea that there is some inevitability underlying social change—be it biologically based or due simply to some motor of change inherent in the very fabric of society. But it is the pattern of change that Bock reports, in an almost offhand manner (as if well known to all), that arrests our attention here: far from being unilinear, gradual, progressive, and inevitable, social change is a rare phenomenon, the result of specific events. When change happens, it happens quickly. For the most part, history is nonchange unless and until something happens, a specific event occurs, that knocks a culture off its accustomed track and forces change to occur. And though we agree with Bock that explanations of cultural change in human history in biological terms is neither illuminating nor compelling, we will not ask why the patterns of biological and cultural change seem so remarkably similar until the next chapter. Here we only seek to establish the pattern of cultural change.

Bock mentions only a few historians and social scientists who support his view of patterns of change in history. But among these was the remarkable F. J. Teggart, a historian who spent the greater part of his career at the University of California, Berkeley, in the first half of this century. Teggart saw the study of human history divided into two basic camps, neither of which seemed to him particularly capable of explaining "how things came to be as they are." On the one hand, there was the actual writing of history—a litany of sequences of events with little care given to what Teggart called the "scientific" study of the causes underlying history's events. Teggart viewed the history of his day as more of a literary exercise, full of speculations on the motives of historical figures, than a serious enquiry into the whys and hows behind the development of cultural diversity as we see it today. Too much preoccupation with "facts" obscures overall pattern.

On the other hand, Teggart saw the other half of social science's approach to cultural change in even less flattering terms. In his brilliantly constructed essay *Theory of History* (1925) Teggart traces the Victorian notion of progress back to Descartes' vision of a universe in constant motion. Social philosophers before and after Darwin tended to see change—slow, steady, progressive change—as inevitable within cultures. To such men as Auguste Comte, historical events were meaningless in the study of the "evolution" of society. What seemed important were all the features common to societies. Cultural evolutionists then (and now) see the differences between societies, those products of different events affecting separate societies, as obscuring the basic, true nature of sociocultural change which is common to all groups. Teggart (rightly, it seems to us) thought both approaches a poor way to understand how things have come to be as they are. One, seeing only events, with vague speculations about causes, was "unscientific," while the other was a theoretical evolutionary approach which had scant use for the events—the actual *data*—of history.

Teggart's dim view of cultural evolution saw problems similar to those we have already met in evolutionary biology. The assumption of slow, steady, progressive change led to a definite expectation of what the evolutionary process was like, how it worked, and what the patterns it produced should look like. As we have seen (and as Teggart, astonishingly, pointed out in 1925) the assumption that biological evolution is a matter of gradual, steady, progressive change flies in the face of the evidence: the actual historical events that biologists are supposed to be explaining with their evolutionary theories. We have already argued that the fossil record flatly fails to substantiate this expectation of finely graded change. So too, says Teggart, does the historical sequence of events in human history.

Well, if the events of history falsify the notion of gradual, progressive change in societies, what views of history are we left with? And how do we go about studying processes underlying historical change?

Teggart was able to trace a third line of thought which seemed to him to fit the real-world situation of history and cultural change better. The view Teggart—and we—see as both a more accurate description of the way things are and a better program of research into cultural change can be traced back at least as far as the Scottish philosopher David Hume. In a nutshell, there are three ingredients to cultural change: stability, slow modification, and the emergence of the "truly new." The proper approach to history should be the comparative study of societies to see how all three factors operate to produce the societies we see today.

Stability, according to Teggart, means that "certain activities—that is to say, ways of doing things and modes of thought—have been maintained with recognizable uniformity from age to age." Imitation, sympathy, habit, social pressure, and the like conspire to create fixity and persistence; and it is clear that, all else being equal, such stability is the primary expectation of the historian.

Next, there *is* a certain tendency to change inherent in all cultures. Teggart's own example was linguistic change, where changes in vocabulary and usage can occur from generation to generation. But such change generally leads nowhere radically new and in any case can rarely be construed as "progress." What, then, leads to the third item, true innovation?

Events—or, as Teggart saw them, intrusions or disruptions—struck him as the stuff of real sociocultural change. Collisions between cultures (recalling Kubrick and Clarke's scenario at the East African water hole so many millions of years ago) seemed to him particularly to underlie radical departures in the histories of societies. When Teggart was a young man, China struck everyone as a monument to historical stability, with essentially the same political organization, economy, and mores (in short, culture) that had been developed thousands of years earlier. Though aspects of China's peculiar brand of feudalism are still readily apparent, the events of the recent past set off by extensive contact with foreign cultures have wrought more change in a single century than accrued over the preceding five thousand years.

Expectations of a pattern of social stasis punctuated by change, of innovations triggered episodically by major, but rare, intrusions (usually from outside) and followed by long periods of little or no change, color what we see—just as the view of gradual progress has distorted many a narrative of the history of the Western world. When we point to China, or Egypt as confirmation of these general ideas of patterns of cultural change in historical times, we freely concede we are seeing history according to a presupposed historical outlook. But at the same time it seems that there are some regular patterns of change in human history, that these patterns can be seen objectively, and that a dominant, if not the sole, major pattern is Teggart's and Bock's "stability punctuated by brief episodes of change." Egypt is a particularly compelling case in point.

Patterns of Change in Egypt

Ramesses II, whose temples and statuary likenesses pepper the entire length of the Nile valley, is renowned as one of history's greatest egomaniacs. But the vainglorious king of Shelley's "Ozymandias" was also one of Egypt's most successful rulers, expanding Egypt's domain and presiding over yet another of many episodes of prosperity.

Sometime in the thirteenth century B.C., Ramesses sent one of his many sons, Kha'emwaset, to explore and restore some of the funerary monuments at the old cemetery of Memphis at Sakkara. Kha'emwaset left an inscription on the south side of the (then and now) rather ruined pyramid of the Fifth Dynasty king Unas, who had lived some 1,100 years earlier. The inscription is one of the earliest records of historical research: Kha'emwaset states his belief (correctly) that the pyramid belonged to the ancient king Unas.

Here we have the first two themes of Egyptian history together: the unification of the two lands into a single Egypt, the pharaonic system itself, the elaborate cemeteries with mortuary temples, all remained pretty much the same from Unas' day down to Ramesses—and were to persist a good deal longer. But there had also been change: Unas' pyramid was one of the last built. When Ramesses was pharaoh, the tradition for the burial of kings and nobility had long since switched from pyramids to long shafts excavated in the cliffs and desert floor along the perimeter of the arable land (though "switched" is too strong a word, as shaft burials were already in use in the Old Kingdom when the pyramids were built). And inside Unas' pyramid, still accessible to tourists, are the earliest known pyramid texts: ancient incantations in old hieroglyphics that were hard indeed for Egyptians of the New Kingdom to understand. Here is an example of the sort of drift—a gradual change in usage—that Teggart had in mind when he spoke of the propensity for *some* change to accrue through time.

So the world of the Old Kingdom was both familiar and a bit strange to its descendants living in the New Kingdom a thousand years later. Historians rightly grumble that the stability of customs and social institutions in Egypt has been emphasized ad nauseam. But the truth is that the world of Kha'emwaset and Ramesses II was a good deal more like Unas' world than anyone might expect given the lapse of 1,000 years. And much of the change that did occur came through intrusion, as when the "shepherd kings" (the mysterious Hyksos) introduced the horse-drawn chariot as they invaded the Delta in the interregnum between the Middle and New Kingdoms.

The stability of Egyptian culture for nearly 3,000 years (some would say 5,000 years: the fellahin of the Egyptian countryside still labor in the fields with many of the same implements used by their ancient counterparts) was pervasive, affecting all aspects of their way of life. Easiest to see is the periodic renaissance of older styles and themes in Egyptian art: the art historian can usually tell the proper period of a sculpture or temple carving without the aid of telltale hieroglyphic inscriptions. But the stability of style and content of the artwork impresses even art historians. The usual explanation: later artists simply repetitiously copied earlier works. Some of the finest art in all of Egypt's long history was produced at the outset, during the Old Kingdom, soon after the unification of Upper and Lower Egypt. In their haste to build more and more splendid monuments, Ramesses' artisans usually spurned the time-consuming older practice of carving reliefs by chiseling the entire background away. Instead, their "incised relief" merely excavated the periphery of the figures—a quick and dirty rendition of a style executed in less hurried times. But later, for instance in the Ptolemaic temple at Kom Ombo (built some 1,000 years after Ramesses) we find again true bas relief, fully as exquisite as examples 2,000 years older.

Egyptian cultural stability extended far deeper than artistic expression in religious and funerary buildings. The basic physical facts of Egyptian existence placed severe limitations on the mode of life of the people inhabiting the delta and the extremely narrow verdant strip along the river between the two deserts. The practical exigencies of coping with the predictable, but yearly changeable, Nile have lain at the heart of Egyptian existence, from the beginnings of permanent settlement and agriculture to the present day—the Aswan High Dam notwithstanding. As the historian Karl Wittfogel and the anthropologist Marvin Harris have stressed, control of the yearly flooding and retention of water into systems of canals and lakes required an integrated political control, necessitating the rapid invention of a centralized government under the command of a single ruler—one of Wittfogel's "oriental despots." Once instituted, the pharaonic system obviously worked rather well.

The two periods separating the Old Kingdom from the Middle and the Middle from the New were times of breakdown of the system. Internal strife, fragmentation of authority along the Nile, and quick turnover of rulers mark these periods, about which little else seems to be known. Recent research has convinced some historians that these periods of decline were correlated with, and probably triggered by, prolonged periods of crop failure—occasioned by too much or (more often) too little flooding of the Nile. Strong as it was, the stability of the Egyptian state under a single pharaoh was ultimately dependent upon the annual "gift of the Nile." In a system geared to provide food and maintain order in lean times, there were clearly limits to the power of even a god-king. But, again and again, good times followed bad, and much the same sociopolitical system, with all its cultural trappings, would once again appear.

Egypt's physical isolation, with desert to the west, south, and east, and the Mediterranean to the north (seafaring came after the initial rise of Egyptian civilization), usually gets the credit for the astonishing lack of change its history reveals, for Egypt's natural protection against foreign incursions is a good test case—in reverse—for Teggart's idea that major historical change reflects periods of disruption coming usually from *outside* a culture. The Egyptians were left alone for centuries on end, and their society changed but little. Their first major invaders, the Hyksos, ruled for a while but effected no major change beyond leaving the horse-drawn chariot. And even Alexander's conquest, which led to Greek control of Egypt under the Ptolemies in the final·three centuries B.C., again showed that Egyptians had a greater influence on their conquerors than vice versa. Only with the Christian era and, more important, the advent of Arab control (A.D. 640) do we find a radical change in the religion, language, architecture, art, and the political system. When substantial change finally did come to Egypt, some 3,500 years after its inception as a nation-state, it came as Teggart would have predicted: as (finally) a capitulation to the force of another culture. But even now the fellahin use the same pots on counterweighted beams to raise water from the river and adjacent canals up to the level of the fields. Some things never did change.

Everyone would agree that Egyptian civilization had no great impact on the subsequent development of civilization and society in Europe: quite obviously, it was a form of economic and socioreligious organization appropriate to a highly specific set of environmental circumstances. Greek civilization, however, plainly laid the foundations for later cultural development in Europe. But is the pattern of change from the Greeks to ourselves one of slow, steady, progressive development? Or do the real cultural changes, in government and family structure, in language and in the arts, reflect instead an episodic tempo?

It seems fairly evident that the gross pattern of historical development in the West has not conformed to the simple model of unilinear, gradual progress. Greek civilization attained its zenith not long after it first appeared. And its subsequent history was no march of progress toward some more advanced state—indeed, maintenance of the status quo and decline mark the later periods of ancient Greek history.

The Romans were certainly no lineal descendants of the Greeks. Developing elsewhere, and liberally borrowing the most attractive elements of Greek culture, Roman civilization cannot rationally be viewed as an improvement over the Greeks, nor can Roman political and cultural institutions be viewed profitably as a simple function of Greek historical evolution. In short, there is no evidence that Greek culture slowly evolved in such a fashion as to end up as Roman civilization. Yet it is largely through the Romans that elements of ancient Greek civilization survive to the present time.

Post-Roman times bring us to a period of history that we can justifiably point to as "our own." And we may well be disposed to think of this history from the Dark Ages to the present as the story of an uphill trudge from benightedness to our present sophisticated manipulation of nature. But we are in danger of being rushed into such ideas of historical "progressiveness" by the fact that our history is crammed with events—events which it is tempting to equate with change—and, in the aggregate, with slow, progressive change. Clearly, however, such political events as wars, coups d'etat, sweeping invasions of Gothic hordes, shifting of political boundaries and so forth, cannot be construed as change in any profound sense. Although they may have changed—or ended—the lives of thousands of individuals, they rarely had more than a temporary impact on the lot of the mass of humanity. Events which have led to real changes in human existence have been rare. Such events there have been, however.

Of these, the outstanding events in Europe were the Renaissance—a period of unprecedented revolution in human awareness and conception of the world—and the Industrial Revolution, an economic event of a magnitude unparalleled except possibly by the introduction, 10,000 years earlier, of settled agriculture, and of which we are still experiencing the repercussions. Where, then, is the expected pattern of slow, steady historical change? It simply doesn't exist—except in the minds of those who cling to the myth that engendered it.

These few cursory remarks about history agree with the patterns seen, for instance, by most of Egypt's historians. But emphasis on the stability of cultural systems *is* generally seen as lack of "progress," and many admirers of ancient Egypt are quick to insist that change *did* occur through time. It is as if pointing to stability is somehow to insult a nation, as if the lack of "progress" so widely believed to be inherent in things is a mark of failure, the stamp of a backward people. But the sociopolitical system that so quickly developed in Egypt in response to the exigencies of a settled, agriculture-based way of life in the Nile valley, was admirably suited to the task; so much so that no "improvements" were called for. "Progress" is often an illusion. Stability, after all, is what most of us want in our social systems. Once again we find ourselves agreeing with Teggart (*Theory of History*, p. 220):

To believe in progress is to adopt a supine attitude toward existence; is to cultivate an enthusiasm for whatever change may bring; is to assume that perfection and happiness lie ahead, whatever may be the course of human action in the present. To restrict belief to the possibility of progress implies recognition of the fact that change may result in destruction as readily as in advancement; implies consciousness of the precariousness of human achievement, as witnessed in the fate of "Nineveh and Tyre." Belief in progress rests upon the assumption that "all is for the best," but wavers between the views (1) that the "natural" activities of men, if freed from artificial restraints, must necessarily lead to a perfect condition of social existence, and (2) that this desirable condition is to be reached only through the regulation of "natural" activities by legislation—based upon intuitive judgments. Belief in the possibility of progress forces upon us the question, "How may this possibility be recognized?"; it leads us to understand that, if human advancement is to be assured, the activities of men must be directed by knowledge.

But the inevitability of progress is not only a false doctrine misleading social scientists: Teggart sees it as a dangerous source of false security. If we assume all will work out for the best, that a better world necessarily lies ahead, we may be in for a nasty surprise. If we realize that once the magnitude of a problem is acknowledged we are already in a better position to cope with it, we have a better chance of seeing that better day in the future.

The current shortages of food and fuel—the dual-headed energy crisis the world now faces—is a case in point. As Marvin Harris points out in his *Cannibals and Kings,* it is possible to see the major events of human history—and particularly those episodes of marked increase in population size—as the outcome of successful bouts with past "energy crises" in the sociopolitical production of the very same foods and fuels we worry about today. To point to the past and merely assume we can do it again would still be supine. But to realize that such crises have occurred in the past, that the present situation is alarmingly serious, may at least get us to remove our collective head from the sand. Teggart's moral conclusion, that we can understand our history and thereby shape our future, is a brave message that most of us seem to find less attractive than a naive simple faith that everything will work out okay in the end.

The view of progress, of course, was not first invented by Europeans in the midst of the Industrial Revolution. But there were probably few peoples at any other place and time in human history who more thoroughly adopted progress as the cornerstone of their world view. Perhaps it is because our present times make it easier for us to doubt that progress truly is inevitable, but in any case, it seems to us that the idea of progress (especially of the inevitable, slow, steady, gradual improvement variety) has distorted our very perception of patterns of change. *Neither* the course of life's evolution nor the checkered historical career of mankind to date fits this picture.

In fact, both biological and historical data fit the picture of progress so badly that we wonder how the notion could have become so popular and have hung on so long (more stasis!—the notion of progress has dominated our thinking for a good 250 years now, contrary to the evidence of our own senses). Social Darwinists, apologists for the status quo where a privileged few enjoyed the fruits of the Industrial Revolution while the masses labored in frequently miserable conditions, saw social conditions as the logical, inevitable outcome of laws of nature: survival of the fittest ("natural selection") put the best (including the theorists) at the top of the heap in the continuous competitive struggle. And the struggle was itself simply a natural process of continual change—"progress." But certainly not all believers in progress were so myopic. Most social reformers also believed in progressive change and improvement; they simply wanted to see more of the populace participate.

The illusion that change is inevitable, though usually slow and gradual, comes more from a misperception of pattern than from any specific social ideology. The most conspicuous elements of any culture are its material goods. The Industrial Revolution saw an explosion of manufactured items. Harnessing of fossil fuels particularly spurred the invention of a host of machines to perform a bewildering array of tasks. And we are still living with change—truly rapid change—in our technological trappings. If anything, the pace of change is accelerating. There is little wonder, in an age where grandparents remember Lindbergh's solo flight, children saw Sputnik and real men on the moon, and grandchildren witness the space shuttle that progress remains a compelling idea. Change, impressive if superficial, more than ever appears to be a dominant aspect of our lives. We have long since come to see it as inevitable. And here is a paradox: why do we think that intrinsically change is slow, steady, and progressive if we live in an age of still accelerating change? The answer seems clear: the real patterns of change in history do not readily support any sort of notion of progressive change. To make the idea work, to insist and be believed that progressive change is inevitable in biological and cultural systems, one had to admit it was *very* slow. Why did Egypt remain pretty much the same for 3,000 years? The answer: because change was awfully slow. That such patterns of slow change (which all would agree are more nearly the norm than the truly rapid technological change we have seen in the past 250 years in Western societies) do not lead to the sorts of major change that has occurred in history is conveniently ignored. Teggart and the others who have attacked the idea of progress are certainly not the only ones to have seen patterns of change as episodic rather than slow, steady, and progressive. The old notion of gradualistic progress is itself partly a compromise between the notion of a constantly changing universe and the observation of stasis. It also affords an explanation of the current scene of technological change as merely a more rapid version of normal processes. Ironically, slow, steady improvement is as much the old explanation of stasis as it is an explanation of change.

CHAPTER EIGHT

Beyond Patterns: Theories of Change

WE HAVE come full circle. We have looked at patterns of change in the anatomies and behaviors of animals and plants. We have examined the details of the physical evolution of our own species. And we have also looked at patterns of cultural change—our ways of doing things, of "being human." All along we have looked for the expected pattern: slow, steady, progressive change. And all along we have found instead a pattern of sporadic change.

We are aware of the irony of this. We have debunked the myth that evolutionary change is gradual and progressive. We have likewise skewered the notion that patterns of cultural change—prehistoric and historic, in simple and complex societies—are gradual and progressive. Both are based on the same preconceptions about inherent change. And it is ironic that the patterns in biological and cultural change that we do see end up still being similar. Are we, too, deceiving ourselves? It almost seems too good to be true, or at any rate a bit improbable, that the similar expected patterns of both cultural and biological evolution would both turn out to be wrong, to be instead something quite different and yet remain basically similar to one another. We admit our own cultural biases may well be at work here, and it is true that notions of sudden bursts of change followed by longer stages of quiescence are invading a wide variety of fields these days, including economics, mathematics, and the philosophy of science.

But it seems to us that it is the evidence itself, events in the history of life and the history of mankind, that destroys the myth of gradual, progressive change. The patterns, in other words, speak for themselves. They are there for all to see. We have not simply invented them to fit a set of preconceived notions.

So it only seems natural to wonder why the patterns of biological and cultural evolution are so similar—even if they aren't the same old familiar patterns we have been taught to expect. Is there some grand process underlying the two? Are they really the same? Can we, after all, reduce cultural evolution to biological terms?

That the patterns produced by biological and cultural evolution are similar is one thing. To say that the processes that created them are therefore the same is another. Sociobiologists (as we saw in the quotation from Bock in the last chapter) see the patterns as similar (gradual, progressive) and conclude that the causes must be the same (progressive adaptive change under the guidance of natural selection). Social theorists have also tended to see that the patterns are the same (gradual, progressive) but have looked for different factors, different motors driving the process of change in the two different systems.

The traditional response of cultural evolutionists to biologists, who, like sociobiologists, seek to reduce cultural evolution to the terms of biological evolution, justifiably emphasizes inheritance. After all, the Darwinian concept of natural selection depends upon genetic inheritance. As we have seen, though Darwin knew nothing of the biochemical basis of heredity, he knew that offspring tend to resemble their parents and that change could accumulate from one generation to the next as natural selection favored those individuals best suited to the environment. Social theorists have been fond of pointing to the "Lamarckian" style of inheritance (or transmission) of cultural things. We learn culture, and if we tend to learn quite a bit from our parents, we also learn from our peers (therefore *their* parents), our teachers, and all the other models we acquire in life. We might even learn from books and (God help us) television. This style of transmission is "Lamarckian" because a cultural "item" may be acquired (either learned or invented) in the course of one's lifetime and transmitted in due course to the next generation. Mutations in germ cells likewise appear in one's lifetime and may be passed along, but adaptive properties of organisms are inherited from parents and cannot be modified genetically in the course of one's lifetime. Biological adaptations are modified as slightly variant versions—differences between individuals in a population—are selected from generation to generation. Just because a pattern (in this case the myth of progressive adaptive change) is held in common, it does not follow that we need one single theory to account for its presence in both cultural and biological evolution.

Likewise we need no single theory to embrace the similar patterns of episodic change we see as common to both systems. We may attribute stability within species to the persistence of niches: nonchange is the direct result of a successful instance of speciation, the species merely continuing to exist so long as its environment remains recognizable to it and no disastrous ecological events befall it. There simply is no need or intrinsic urge for a species to evolve itself out of existence. So, too, with social change: once invented, the pharaonic political system of Egypt persisted simply because it worked. Integrative control of the waterways was achieved by concentrating authority under a single god-king. And so long as the Nile behaved and outside disruptions were minimized, little in the way of change could be expected.

But this idea that once a way of being is invented it will persist until change is forced upon the system is hardly a metatheory of evolution. Nor do we think a single, underlying comprehensive theory that explains both cultural and biological evolution, especially one which reduces the former to the latter, is possible. The systems are too different. In addition to the different modes of inheritance (important particularly to a theory of adaptive change, but still relevant here), on a broader scale we are simply dealing with radically different kinds of entities. We need only to look at ourselves to see that this is true.

Today we are but a single species, *Homo sapiens*, and some 4 billion of us have encircled the globe. We are *eurytopic*: our adaptations are broad and general. Our cultures, diverse as they are, serve to fit us to the physical exigencies of the wide variety of environments in which we live. But we are a single species.

The biological theory of change we have presented in this book sees species as basic units in evolution: species are the ancestors and descendants of the evolutionary process. The pattern of change in the fossil record strongly suggests that most change in the evolutionary process is related to the origins of new species. To be sure, species vary within themselves. Our species is highly polytypic: physically and genetically we are very diverse, just as we are culturally diverse. But the pattern of cultural diversity does not correspond particularly closely to the pattern of phenotypic and genetic variation—and social theorists long ago abandoned the idea that cultural differentiation could be readily explained as a by-product of the physical differentiation of mankind.

That patterns of genetic and cultural differentiation are not coextensive should be enough right there to deter biological determinists (like sociobiologists) who seek to reduce cultural evolution to biological terms. But there is more to it: though biological species vary within themselves, speciation (the origin of new species from old) is *not* a simple process of adaptation. It is a different sort of affair, involving the fragmentation of one reproductive community into two or more. It is this process, as we have seen, that encourages the establishment of new behaviors and anatomies in evolution. Without this sundering of reproductive communities, it seems, natural selection on its own would not generally produce much significant change. Speciation seems to be the key to the truly new, the origin of anatomical and behavioral diversity in the biological evolutionary process.

We have seen that the evolution of our own lineage over the

past 4 million years or so is no easy matter of interpretation. Legitimate disagreement on the crucial issue of "how many species" were there at any one time beclouds the issue. But enough is clear: the archaeological record quickly becomes out of phase with the pattern of appearance and disappearance of the various physical types we call new "species" within our own lineage. Today, within our own single biological species, we can find remnants of most of the major tool traditions we see in the archaeological record: what took a lineage with perhaps as many as a half-dozen species to produce (in cultural diversity) we have retained today within a single species—not to mention the explosion in cultural traditions we have seen in the past 35,000 years, when modern man first shows up in the archaeological and paleontological record. If there is any substance to the newer views of biological evolution that we have recounted here, biology has less relevance for understanding human cultural diversity and history than it ever did. Sociobiology is an astonishing anachronism. Our physical and cultural characters simply do not match up, either in space or through time. Our physical and cultural histories are out of phase, and there is no simple correlation between physical and cultural diversity. The processes underlying these patterns of diversity and change are simply independent of one another.

To look for a common theory to explain patterns of both biological and cultural evolutionary change would be akin to the sort of "science" of explaining the evolution of the nonexistent group of "sharks + ichthyosaurs + porpoises" we discussed in chapter 2. The group is not "natural." Evolution did not form it, so it is fallacious to speak of its "evolution." But common principles of hydrodynamics do emerge when these organisms are compared, and in that spirit of limited enquiry, we note again that patterns of cultural and biological evolution are hauntingly similar. And though the processes of cultural and biological evolution cannot be the same, nonetheless there may be some points in common between them. Recalling for a moment Teggart's insistence that it is events that give spice to historical change, we find it equally true that disruptions cause the extinction of species and the fragmentation of old species into new descendants. What sorts of "events" are these?

Conventional wisdom (and in this instance we see no cause to demur) views the fragmentation of species essentially as accidental. A species' distribution is disrupted by shifting patterns of ecological change. An extensive woodland breaks up into prairie land with isolated woody patches as the climate dries out. Or the Pacific and Caribbean are cut off as the Panamanian connection between North and South America emerges as dry land. Extinction and evolution appear as reflections of a changing physical environment. Classic Darwinian theory sees this response as a direct adaptive reaction: change the environment and the species change to keep pace. But the sort of reaction we have in mind here is *geographic*. Climatic change transforms the distributions of environments, often fragmenting once continuous habitat belts. Evolution and extinction largely mirror changes in physical geography. Species are fragmented and thus new ones are born. And species are extirpated if they are unable to relocate as their accustomed habitat moves.

Geography plays a similarly basic role in the ideas of scholars who, like Teggart, see history as an episodic interplay of stasis (with a slow, negligible, intrinsic change) and events. And events are most commonly triggered by changes in the distributions of cultures. Cultural collision, be it armed conflict or the borrowing of an idea, lies at the heart of major historical change. Again, geography seems to play a major role in the "events" that amount to major episodes of change.

Here, then, is a similarity between cultural and biological evolution worth noting: geography seems to play an important role in each. Geography *isolates*. Both cultural and biological evolutionists have for over a century attributed the diversity in the systems they study primarily to the isolation afforded by far-flung geographic distributions. The histories of particular societies or species are, of course, independent of other societies or species with which they have no contact.

But isolation works in other ways, too. On the one hand, there is the opportunity to go in your own direction, to develop your own independent tradition. But isolation can also lead to stasis. Here the effects of geography on cultural and biological evolution become rather different from one another. If Teggart is right when he says that the societies that remain the most consistently isolated will show the most stasis, it is because it takes interactions with other cultures to force change. But, though there is some theory (and substantiating evidence) that interactions between species may be an impetus for change in biological evolution, no evolutionist insists that such interactions are the only force underlying change. Thus, isolation of species may lead to stasis but is not necessary for stasis to occur. Species are more accustomed to (and more successful at) living side by side in well-integrated ecosystems with patterns of mutual interdependence than, as a rule, societies have proven to be.

Thus geographic patterns—especially isolation—can foster change and consequent stability in both evolutionary systems. Why the change need be rapid, as experience shows it nearly always is in both systems, is another question, though "sink or swim" does suggest itself for biological evolution. If change is necessary to make a way of life possible in a given habitat, a small fragment of an ancestral species may be expected theoretically to be best suited to a rapid development of such modified adaptations. Perhaps the same is true for cultural innovation in newly isolated social groups.

The similarity of the response to geographic imperatives can perhaps be pushed too far for the two systems, and certainly the analogy is quickly strained as we think about *how* geography affects change in both cultural and biological evolution. But the overall message is clear: though the processes of cultural and biological evolution are both separate and different, the historical patterns each creates are at least superficially similar—posing a paradox as yet incompletely resolved. And the role of geography in fostering diversity and perhaps stasis in both systems cannot be ignored.

Epilogue: Our Evolutionary Future?

WE DO NOT pretend to be fortune tellers, but the patterns of change we have discussed in this book have strikingly different implications about our future than do the visions ordinarily reported by futurologists.

The idea that change is inevitable tells us the future will be different from today. The remoter the future, the greater the difference will be—just as the farther we go back in time, the more different from us our ancestors seem to have been. And to some extent this all seems reasonable. But how much change, affecting what aspects of humanity, is it reasonable to expect? And how will that change occur?

The best way to predict the weather, it is sometimes said, is simply to call for today's weather to be here tomorrow. You will be right more often than not. If our recent past is any guide, Western society will become even more dependent upon technology for the production of goods and services. There is no reason to expect such growth to stop: technological change should continue to accelerate. But even here it is possible to see patterns of innovation, rapid proliferation, and stasis until more sudden change. The rapid growth of computers has been a function of successive waves of electronic inventiveness. The original UNIVAC took up most of a floor of a block-square building. Its vacuum tubes gave way to transistors and these in turn to miniaturized "chips." All this has happened in thirty years, and those in the industry speak of "generations." The

UNIVAC seems positively paleolithic now. Yet there *is* some stasis, because older generation computers are still in use around the world. Not everyone has the latest model. The longevity of computer "generations" plots out just like the archaeological record of different tool traditions from a bygone era when the pace of change was less hurried. Computers symbolize our modernity, our advanced sophistication. Yet their "evolution" has hardly been linear. No matter how fast a pace technological change achieves, it too is locked into an episodic pattern.

Apart from technology, and barring Armageddon or our own lineage's first failure to deal with a food/fuel crisis (which, of course, could happen; after all, the vast majority of all the species that ever lived are now extinct), what sorts of changes are in store? Turning to the various theories of how change occurs, fans of the "change is inevitable" viewpoint will, of course, talk of all sorts of transformations. Teggart (in the passage cited in chapter 7) and Marvin Harris (in *Cannibals and Kings*) hold out the optimistic view that if we do achieve a rational understanding, a viable theory, of how social change occurs, there is a chance we can control our own destiny. There is really no precedent for this, of course, and therefore distressingly little reason to see it as more than a hope, a wistful possibility.

But if things are left to run their own course, what then? If social evolution is inevitable, we should still expect change. But if the essence of cultural change is stasis punctuated by brief episodes of innovation, can we reasonably expect much to happen given the near standing-room-only condition of the world? We will see continued "evolution" of the argot within most of the world's languages. What is the probability, though, of seeing a new language evolve? New languages, or for that matter, anything culturally new, have to have the luxury of space—isolation—to develop. (We are thinking of true languages, of course; we can invent languages, such as COBOL, FORTRAN, and Esperanto—but who speaks Esperanto?) And the pattern today is definitely away from Kubrick and Clarke's collision of small bands of hunting-gathering australopithecines. Global communication has accelerated the collisions of cultures. Apart from pidgin English and a few other hybrid cultural forms that have arisen, cultural annihilation rather than innovation has been the main result. Innovation in cultural evolution lies in the origin of diversity, and this in turn depends upon the opportunity for cultures to go their own way, have their separate histories, and occasionally collide. We no longer seem to have the right mix of conditions for innovation to occur. Apart from technology, boring stasis and uniformity seem to lie in the tea leaves.

If this scenario of the lack of innovation in all but the technological sphere of human cultural activity is on the right track, the same is even more probable for our biological evolution. Genetic variability will not be depleted appreciably by interbreeding among peoples once isolated from one another. And although we are one big happy interbreeding species, gene flow is remarkably slow among 4 billion people and there is little danger that we will all blend into one monotonously mongrelized mass. Yet, on the other hand, it is clear that there is precious little chance of anything *major* happening in our physical evolution. How are new adaptations going to arise in a breeding pool of 4 billion increasingly mobile people? They aren't. No fear: we will not become legless slugs with huge brainy heads in the foreseeable future.

Older arguments about our future physical evolution were always firmly entrenched in the gradualistic, progressive notion of adaptive change via natural selection. Some worried that medicine, increasing our longevity by giving eyeglasses to the myopic and teeth to the toothless, inadvertently increases the pool of bad genetic variance in our species. Other geneticists, discounting as illusory the negative effects inadvertently accrued from medicine, nonetheless consider undeniable the power of natural selection to wreak real, significant change in the gene frequencies of the world's people. The great geneticist Theodosius Dobzhansky, for example, thought it the height of benighted self-centeredness to think that we were through with our physical evolution—as if we had reached a pinnacle and could safely ignore our animal heritage, having finally finished with our physical evolution and switched over to the cultural mode.

But Dobzhansky's view was firmly tied to the notion that evolution boils down to genetic change. Given enough time, significant amounts of genetic change are bound to accumulate, even in species with such enormous populations as our own. And time there will be. But Dobzhansky also knew a great deal about speciation. And it seems certain he would have agreed that the possibilities of a speciation event within our lineage in the foreseeable future are two: slim and none. And, if this book's biological thesis is correct, most major change is related to speciation events. Certainly no one has ever shown much real evolutionary change to occur in lineages where there has been little or no speciation.

And we need geographic isolation to fragment an old species, to give us two or more reproductive communities where once there was but one. Space ships founding colonies on Mars or in space stations won't do it because of the lifeline of support required. And ours *is* a eurytopic species—a generalist, jack-of-all-trades kind of species with a broad ecological niche. Such species are hard to break up into permanently fragmented reproductive communities. Even prolonged periods of isolation won't necessarily suffice, as perhaps witness the Tasaday people of the Philippines, "discovered" only recently (though long known to themselves) but still very much *Homo sapiens*. Tough to accomplish in any case, fragmentation of *Homo sapiens* into two or more descendant species is a ridiculously remote possibility. So is the prospect for any significant change in our physical beings.

There is, of course, a grim rejoinder. Ecosystems are constantly being "downgraded" to an earlier stage of succession. The ecologist R. J. Johnson wrote of a sandbar that slowly migrated across the floor of Tomales Bay in northern California for several seasons, wiping out well-entrenched communities of bottom-dwelling marine life. As the sandbar moved along and once again flat sea bottom was exposed behind it, the "normal" marine community of the bay had to start all over again the sequence of events that finally lead to the presence of a mature, stable community.

So it could be with us. Some catastrophe, presumably of our own making, could free up the space, as it were, vastly diminishing our 4 billion to some smaller number. We could once again be distributed as in the old days, which could perhaps lead to change after a binge of the "New Dark Ages." What price change then? Who *needs* it? Who wants it? Would that Teggart and Harris were right: that we are finally in a position of self-understanding sufficiently great to take conscious self-control of our own collective lives. Now *that* would be an innovation.

Index

Acheulean hand-axe culture, 11, 94, 100, 145, 150, 154, 157
Adaptation, 177, 180, 195; by natural selection, 25, 41–43, 44–45, 48–49, 51, 52, 62–65; *see also* Change, constant, adaptive
Adaptive landscape (concept), 40–43, 44
Adaptive radiation, 59–60, 61, 63
Africa, 7, 8, 86–117; fossils in, 80–83, 120, 121, 124, 149–50; as site of early *Homo*, 72, 94–95, 122–23, 136, 140–41; toolmaking traditions in, 10–11
African Genesis (Ardrey), 23
Allometry, 132
Altamira, 159
Ancestry, 42, 117, 123, 125, 127
Ancestry/descent, 25, 31, 35–36, 38, 48, 49, 57, 121, 127–28; speciation is process of, 52–53, 62, 64, 178, *see also* Evolutionary relationships
Andersson, J. G. 83–84
Andrews, Peter, 130
Anthropology, 13, 22
Apes, 17, 26, 69, 72, 74, 121; differences/similarities with man, 27; hominid differentiation from, 129–31
Arago (site), 149, 150
Arambourg, Camille, 101, 110
Archaeological record, 3, 6; *see also* Fossil record
Ardrey, Robert, 91–92; *African Genesis*, 23
Aristotle, 19, 30

Art, 4, 159; Egyptian, 166–67
Artiodactyls, 56
Asia, 8, 153
Australopithecus (australopithecines), 86–87, 88, 89, 91, 94–95, 136; gave rise to *Paranthropus*, 126; gracile form, 90, 95, 105–6, 109, 120, 121, 123, 124–45; robust form, 95, 96, 104–6, 109, 120, 121, 122, 123, 124–25, 157
Australopithecus afarensis, 116–17, 126, 131–35, 156
Australopithecus africanus, 23, 80–83, 97, 107, 116, 117, 123, 129, 131–36, 137–39; dating of, 124, 126; different species from *Homo erectus*, 146; gracile form, 93, 125; skull form, 142
Australopithecus boisei, 105, 107, 117, 157
Australopithecus prometheus, 91
Australopithecus robustus, 93, 117, 125, 133, 157
Australopithecus transvaalensis, 87

Behavior, 4, 6, 16, 24
Bilzingsleben (site), 149
Biological change, 162; patterns of, similar to cultural change, 4–5, 6, 8, 11–12, 22–23, 175–81; *see also* Change; Evolution
Biological determinism, 178
Biological evolution, 6–8; cultural evolution reduced to, 4–5, 11–12,

Biological evolution (*Continued*)
14–15, 22–23, 161–73, 176–79; future of, 185–87; *see also* Evolution
Biology, 13, 16; evolutionary, 2–5, 34 (*see also* Sociobiology); myths in, 43, 44
Bipedalism, 7, 96, 111–12, 131, 135, 144; hominid footprints, Laetoli, 115–16; *see also* Upright posture
Black, Davidson, 84, 85
Bock, Kenneth, 165, 176; *Human Nature and History*, 161–62
Bodo D'Ar (site), 149, 150
Body size, 159
Bohlin, Birger, 84
Boise, Charles, 95
Bone structure, 147
Boucher de Crèvecœur de Perthes, Jacques, 67–68, 71
Boule, Marcellin, 76, 77, 115
Brain, C. K., 89–90
Braincases, 81, 159; *A. africanus*, 135; East Turkana, 104; *Homo erectus*, 142; Java man, 74; Neanderthal man, 151
Brain size, 7, 8, 78, 79, 81, 123, 124; *A. afarensis*, 133; *A. africanus*, 135; East Turkana skulls, 109; ER-406 skull, 104; ER-1470 skull, 106–7; European hominid fossils, 148–49; *Homo erectus*, 143; *Homo habilis*, 137; increase in, 159; Kabwe skull, 150; and language ability, 158–59; Neanderthal man, 69 151; Olduvai Gorge hominids, 95, 96–97; and species classification, 137–39, 140, 143–44; and upright posture, 90–91, 115–16
Breeding groups, *see* Reproductive communities
Broom, Robert, 86–88, 90–91, 92, 93
Browridges, 82, 97, 135, 147, 153
Bruckner, Eduard, 76–77
Burials, 71, 151
Busk, George, 68, 70

Cannibals and Kings (Harris), 2, 171, 184
Cartmill, Matt, 126
Cave paintings, 159
Chalicotheres, 56
Chance, 62–63, 64–65
Change, 11, 50, 51, 128, 162, 166; constant adaptive, 2, 6, 7, 29–53, 57–58, 62, 65, 108, 120, 127, 156, 159, 162, 163–64, 169–70, 171–73, 175–76, 177, 185; episodic, sporadic, 3, 4, 6–7, 9, 12, 38, 65, 161–73, 175, 177, 180, 181; in fossil record, 55–65; generational, 35, 39, 40, 43; historical, 180–81; inevitability of, 4, 65; patterns of, in biological and cultural evolution, 4–5, 6, 8, 11–12, 22–23, 175–81; produced in new species, 61–65; rapid, abrupt, 38, 65, 161–73; technological, 4, 9, 10, 11–12, 180–81; theories of, 175–81; *see also* Cultural change; Evolution
Chapelle-aux-Saints, La (France), Old Man of, 70, 75–76
Characteristics: acquired, inheritance of, 4–5; commonalities of, 17–18, 24–25, 26–27, 31, 57; new, 33
Chariots of the Gods (von Daniken), 23
Chatelperronian toolmaking industry, 154
China, 83–86, 131, 145, 150; *Homo erectus*, 143, 144; stasis/change in, 164–65; *see also* Peking man
Choukoutien, China, 83–86, 89, 97, 142, 143; archetype of species *Homo erectus* in, 144–45
Civilization, Western, 162, 168–69, 183
Clark, Wilfrid Le Gros, 90, 115–16
Clarke, Arthur C., 6, 10, 124, 164, 184
Climate, 130, 142, 145, 148, 180; fluctuations in, 76–77; Olduvai Gorge, 99–100; and reproductive isolation, 155–56
Competition, 6, 159; for resources, 33
Comte, Auguste, 163
Cooke, Basil, 108, 109
Coppens, Yves, 110
Creationism, 20–21
Cretaceous period, 59, 61
Cro-Magnon, France, 71
Cro-Magnons, 77
Cultural change, 6, 32, 33; future of,

184–85; ingredients of, 164–65; patterns of, 162–73; similar to biological change, 4–5, 6, 8, 11–12, 22–23, 175–81; tendency to, 164, 166
Cultural collision, 180–81, 185
Cultural evolution, 4–5, 22; reduced to biological evolution, 11–12, 14–15, 22–23, 161–73, 176–79
Curtis, Garniss, 97, 108
Cuvier, Georges, 68

Dali (site), 150
Dance of the Tiger (Kurtén), 21–22
Daniken, Erich von, *Chariots of the Gods*, 23
Dart, Raymond, 80–83, 86–87, 91
Darwin, Charles, 30–36, 42, 58, 69, 78–79; *Descent of Man*, 72; descent with modification, 44, 45–46, 48, 52, 53; *On the Origin of Species*, 22, 32–33, 35, 38, 67
Darwinism, 71, 176
Dating, 76–77, 79–80, 88–89, 92–93, 157, 158; absolute, 108; Africa, 124, 125–26; *A. afarensis*, 117; East Turkana, 107–10; Hadar, 111, 112, 113; Laetoli, 114; Olduvai Gorge, 94, 97–100, 123; Omo River deposits, 102–3; skulls, ER-1470, 107–9
Demosthenes, 19
Descartes, René, 163
Descent, 25; with modification, 30–31, 34–35, 38, 44, 52, 65, *see also* Ancestry/descent
Descent of Man (Darwin), 72
Descent of Woman, The (E. Morgan), 23
Determinism, 62–63, 64–65, 178
De Vries, Hugo, 38–39
Dimorphism: *see* Sexual dimorphism; Size dimorphism
Dinosaurs, 55, 59
Diversity, 42–43, 50, 51, 60, 179; among hominid populations, 155; in *Homo sapiens*, 178; and innovation, 185; isolation and, 180
Djebel Irhoud (site), 152
Djebel Qafzeh (site), 154

Dobzhansky, Theodosius, 42, 44, 185–86
Dubois, Eugene, 72–75, 85, 124, 139, 143

East Turkana (site), 103–10, 122, 123, 125, 126, 137, 138–39; *Homo erectus*, 142, 143, 144; robust hominids, 141; stone tools, 140, 145
Ecological crises, 58–59, *see also* Environment
Ecological niches, 41, 55–60, 61, 177, 186; change in, 42–43; of hominids, 120, 122
Ecology, 130, 182
Ecosystems, 59, 60–61; downgraded, 186–87
Egypt, patterns of change in, 165–69, 170, 173, 177
Einstein, Albert, 12
Energy crises, 171
Environment, 8, 130–31, 135, 177; and change, 33–34, 41–43, 49, 50, 53; and extinction and evolution, 180–81; fragmentation of, 142
Eoanthropus dawsoni, 79, 80
Eocene epoch, 43, 55, 56
Eohippus (dawn horse; *Hyracotherium*), 55–56, 62
Erect posture, *see* Upright posture
Eukaryotic cells, 15, 27
Europe: hominids in, 148–49, 159; modern man in, 159, 159; toolmaking industries, 154
Events, disruptive: cause extinction and fragmentation of species, 179–81, 186, 187; trigger change, 164–65, 169–70
Evernden, Jack, 97
Evolution, 2, 5–12, 17–18, 126–28; between/within species, 62–64; decoupling of technological advance from morphological innovation in, 136; governed by ecological rules, 61; large effects of, 38, 39; is matter of ancestry and descent, 25–26; myth of constant adaptive change, 2, 3, 4, 6–7, 29–53, 57–58, 62, 65, 108, 115–16,

Evolution (*Continued*)
  120, 127, 156, 159, 162, 163–64, 169–70, 171–73, 175–76, 177, 185; pattern in, 119–59; as read in fossil record, 126–28 (*see also* Fossil record, interpretation of); reductionism in, 13, 14; synthetic theory of, 37–40, 41, 42, 43–44, 48–53, *see also* Biological evolution; Cultural evolution
Evolutionary novelties, 26, 127, 128; *see also* Innovation
Evolutionary relationships, 57, 108, 121, 127–28
Evolutionism, 22–23; and Creationism, 20–21; extinction and fragmentation in, 180; modern, 34–37; proved in fossil record, 57
Evolutionary trends, 64
Extinction, 58–59, 60, 63, 68, 124–25, 151–52, 184; causes of, 180

Ferrassie, La (France), 70
Fire, use of, 85, 91, 145
Fission-track dating, 113
Fluorine dating, 79–80
Fossil record, 7–8, 34, 81–82; chronology of, 117, 121–22, 128–59; fraud in, 78–80; gaps in, 37, 40, 45–46, 58–61, 127, 128–29; hominid, 67–117; interpretation of, 126–28, 131–32; new species in, 178–79; pattern of change in, 55–65, 163; recognition of species in, 119–20; species as entities in, 45–48, 53
Frere, John, 67
Future, human, 183–87; can be shaped by man, 171, 184, 187

Galley Hill, England, 78, 80
Genealogy, 31, 34–37; *see also* Ancestry/descent
Genetic drift, 62–63
Genetics, 14, 16, 17, 37–40, 43; population, 35, 39
Geographic isolation, 50–52, 186
Geography, 130; of evolution, 8, 180–81

Gibraltar, 68, 70
Gigantopithecus, 129, 131
Glaciations, 76–77, 155
Greek civilization, 168–69
Growth, limit to, 33
Günz (glacial), 77

Hadar (site), 110–14, 116, 117, 125–26, 131–32, 133, 136
Haeckel, Ernst, 72, 74, 75
Haile Selassie, 101
Harris, Marvin, 167, 187; *Cannibals and Kings*, 2, 171, 184
Heidelberg man, 77–78
History, human, 3–4; patterns of change in, 161–73, 180
Holocene epoch, 77
Hominidae, 7, 26
Hominoidea, 26, 129–30
Hominid footprints (Laetoli), 7, 114–16
Hominid species, 7–8, 77, 155; abrupt displacement of, 151–52; coexistence of, 138–39, 140–41, 154; differentiation of, from apes, 129–31; duration of, 8; in fossil record, 67–117; fragmentation and isolation of, 155–56 (*see also* Reproductive isolation); gracile form, 137, 140; lifestyle of, 135, 151, 156; lineages of, 64, 117, 121–28, 133, 140–41; new, 7–8; nomenclature of, 93, 125; robust form, 76, 77, 140–41, 143, 144, 151; single-species hypothesis re, 117, 120–22, 126, 127; three-species hypothesis re, 122–24, 126; two-species hypothesis re, 124–26; variability of, 144–45, 185–86; *see also* Dating; Fossil record; Speciation; Species
Homo, 16, 17, 126, 133, 141; question of what to include as, 26–27, 125, 136–41, 146
Homo erectus, 75, 86, 97, 106, 109, 117, 121, 122, 123, 125, 133, 137, 141–46, 148, 150, 151, 156, 157; ancestral to *Homo sapiens*, 142–43
Homo habilis, 96–97, 103, 108, 109, 117, 122, 123–24, 125, 133, 136–40,

157; ER-1470, 107; validity of, 126, 139–41, 142
Homo neanderthalensis, 69
Homo sapiens, 7, 15, 17, 23, 26, 117, 133, 137, 158; antecedents of, 97; archaic, 146, 150, 151–54, 155; eurytopic species, 178, 186; *Homo erectus* ancestral to, 142–43; myths re origin of, 1, 4; possibility of speciation, 186–87; is single species, 178
Hooton, Earnest, 90
Horse evolution, 43, 55–57, 62, 64
Howell, Clark, 101, 103, 110, 134
Howells, W. W., 147, 155
Human Nature and History (Bock), 161–62
Hume, David, 164
Hunting, 91, 151, 154, 158, 159; by Olduvai hominids, 100
Huxley, Thomas Henry, 69
Hyksos, 166, 168

Ibn Khaldun, 19, 26
Ideas, change in, 65
Implements, *see* Tool making and use
Industrial Revolution, 2, 30, 170, 171, 172
Inheritance, 4, 33, 38–39, 51; of acquired characteristics, 4–5; and cultural evolution, 176–77; of viciousness, 92
Innovation, 164–65; in cultural evolution, 185; in technological change, 183–84; *see also* Change
Interglacials, 77, 149
Interstadials, 77
Intrusions: trigger change, 168; *see also* Events, disruptive
Inventions, 6, 8, 10
Isolation, 180–81; *see also* Geographic isolation; Reproductive isolation

Java man, 72–75, 77, 78, 85, 95, 97, 120, 124, 139, 143–44, 150
Jaws: East Turkana, 106; Hadar, 112, 131–32; Laetoli, 114, 131–32; Mauer, 148; Omo River deposits, 103, 134

Johanson, Donald, 110, 112, 116–17, 126, 131, 132, 133; *Lucy*, 117
Johnson, R. J., 186

Kabuh Formation, 143
Kabwe (site), 149, 150
Kanam (site), 93
Kanapoi (site), 117
Kanjera (site), 93
Karari stoneworking industry, 109–10, 145
Kay Behrensmeyer Site, 104
KBS tuff, 104, 107, 123, 141; stoneworking industry, 109–10
Keith, Arthur, 75, 78, 79, 90; *New Discoveries Relating to the Antiquity of Man*, 87
Kenyan Omo Research Expedition, 101–3
Kha'emwaset (son of Ramesses II), 165–66
King, William, 69
Klein, Richard, 159
Kohl-Larsen, Lars, 114
Koobi Fora (site), 104, 109, 126, 138; Research Project, 107–8
Kromdraai (site), 88–89, 91, 93, 140–41
Kubrick, Stanley, 6, 10, 124, 164, 184
Kuhn, Thomas, 65
Kurtén, Björn; *Dance of the Tiger*, 21–22

Laetoli (site), 114–17, 125, 131–32, 150; hominid footprints, 7, 114–16
Lamarckianism, 176–77
Language, 29, 158–59; change in, 164; new, 184–85
Lantian (site), 144
Lascaux, 159
Leakey, Louis, 93–101, 114, 122–23, 136, 142
Leakey, Mary, 94–95, 99, 114, 116
Leakey, Richard, 101, 103–4, 107-8, 123, 142
Learning, 4
Life: classification of, 35; diversity is pattern of, 42–43, 60, 127;

Life (*Continued*)
evolutionary history of, 35, 40, 41, 42, 43, 44; fit of hominid fossil record with history of, 128–56; hierarchy of similarity in forms of, 30–31
Linnaeus, Carolus, 30–31, 36, 58
Lothagam (site), 117
Lovejoy, Owen, 116
Lower Tertiary (period), 61
Lucy (skeleton), 112, 113, 116, 132
Lucy (Johanson), 117

Makapansgat (site), 91, 93, 124, 134, 135
Malthus, Thomas, 32–33
Mammals, 24, 55–56, 61
Man: early, 99–100 (*see also* Hominid species); place of, in nature, 17–18, 19–27
Man, modern, 19, 120, 123, 146, 179; abrupt appearance of, 155, 158, 159; ancestors of, 21–22, 26, 27, 72–73, 74, 77, 78; bone structure of, 147; brain size of, 144; in Europe, 154; uniqueness of, 8–9, 17, 18, 158; variability of, 147–48
Mankind, study of, 12–18, 21
Material culture, 9–12, 120, 172; *see also* Stone tools
Mauer, Germany, 77
Mayer, Friedrich, 69, 70, 76
Mayr, Ernst, 44, 45
Mendel, Gregor, 38
Mesozoic era, 55, 58, 61
Middle Stone Age, 11, 154
Mindel (glacial), 77, 149
Miocene epoch, 129–30
Missing link(s), 34, 58, 72–73, 82
Morgan, Elaine, *The Descent of Woman*, 23
Morgan, Lewis, 22
Morphological space, 146
Morphology, 108, 119, 121, 132, 136; of *A. africanus*, 135; of invading hominids, 152; of Neanderthal man, 68, 76
Mousterian stoneworking tradition, 75, 151, 154

Mutation(s), 38–39, 177
Myth(s): of constant adaptive change, 2, 6, 7, 29–53, 57–58, 62, 65, 108, 120, 127, 156, 159, 162, 163–64, 169–70, 171–73, 175–76, 177, 185; contemporary, 1–5; re man's place in nature, 19–27; paleontological, 126–27, 145–46

Natural selection, 25, 33–34, 35, 38, 172, 176, 178; adaptation through, 25, 41–43, 44–45, 48–49, 51, 52, 62–65; and generational change, 39–40; and genetic variability, 185–86
Nature: economy of, 51; laws of, 172; man's place in, 17–18, 19–27; pattern in, 25, 30–31, 62, 63, 65
Nature (journal), 83, 84, 90
Ndutu (site), 143
Neander valley, 68
Neanderthal man, 21–22, 67–72, 76, 78, 95, 115, 120, 121, 151–54, 155, 158, 159; coexistence of, with modern man, 154; dating of, 77; longevity of, 155–56; toolmaking by, 154
Neo-Darwinism, 42, 44, 51
New Discoveries Relating to the Antiquity of Man (Keith), 87
Ngandong faunal zone, 150

Oakley, Kenneth, 79–80
Oldowan stoneworking tradition, 10–11, 94, 95, 99–100, 109, 136, 145, 157
Olduvai Gorge, 93–100, 109, 114, 115, 122, 123, 124, 125, 135, 136; *Homo erectus*, 142, 143, 144; *Homo habilis*, 139; robust hominids, 141; stoneworking industry, 10–11, 109–10, 140, 145
Omo Basin deposits, 100–3, 131, 134, 135, 136; Kibish Formation, 152; robust hominids, 141; stone tools, 140
On the Origin of Species (Darwin), 22, 32–33, 35, 38, 67
"Osteodontokeratic" culture, 91

Paleoanthropology, 87, 93, 126, 137; myths in, 126–27, 145–46
Paleontology/paleontologists, 34–35, 36–37, 67, 71–72; and evolutionary theory, 37, 39–40, myth in, 126–27; and species as entities, 45–46
Paleozoic era, 58
Paranthropus, 125, 126, 133
Paranthropus boisei, 125, 141
Paranthropus robustus, 88, 91, 93, 125, 140, 141
Paranthropus stage, 122
Pattern, misperception of, 172–73
Patterns: in fossil record, 55–65; in history, 161–73, 180; in human evolution, 11–12, 119–59; of similarity and diversity, 2, 175–81
Pei, W. C., 85
Peking man, 83–86, 95, 97, 157
Penck, Albrecht, 76–77
Perissodactyls, 56
Petralona (site), 148, 149, 150, 157
Physics, 12–17
Pilbeam, David, 130, 155
Piltdown man (hoax), 78–80
Pithecanthropus alalus, 72–73, 85
Pithecanthropus erectus, 74, 124
Pleistocene epoch, 74, 76–77, 78, 94, 97, 109, 125, 150; *Homo erectus* in, 141–42
Pliocene epoch, 78, 103, 109
Plio-Pleistocene epochs, 120, 129; coexistence of hominid fossils in, 122; dating of, 102–3
Pope, Alexander, 21
Potassium-argon dating, 98, 102, 104, 108, 112, 113
Prehistory, 4, 9, 161
Primates, 24, 26; *see also* Apes
Progress, 169; evolution as, 29, 30, 32, 33–34, 48–49; inevitability of, 170–72, 183, 184–85 (*see also* Change, constant adaptive); myth of, 2–4; notion of, 30, 32, 163, 164; in technology, 9–12
Prunier-Bey, Dr., 69

Quina, La (France), 70

Ramapithecus (ramapithecines), 129–31
Ramesses II, 165–66, 167
Reductionism, 12–18; of cultural evolution to biological evolution, 11–12, 14–15, 22–23, 161–73, 176–79; in evolutionary theory, 43–45, 51, 65; in sociobiology, 24–25, 26
Religion, 20–21
Reproduction, sexual, 24, 35, 49
Reproductive communities, 8, 35, 42, 186; fragmentation of, 178; species as, 36–37, 45, 46–47, 49–50
Reproductive isolation, 142; and speciation, 49–52, 59
Reproductive success, 33, 62, 64
Riss (glacial), 77, 149
Robusticity, 88, 91, 95, 104–5; loss of, in modern man, 155, 158; *see also Australopithecus robustus*; Hominid species, robust form
Rockefeller Foundation, 84, 85
Roman civilization, 169

Sacher, George, 158–59
St. Césaire (site), 154
Saldanha (site), 150
Schoetensack, Otto, 77–78, 148
Schwalbe, Gustav, 76, 77, 120
Science, 1–2; reductionism in, 12–17
Sedimentary rock, 73, 88, 98–99; in Omo River deposits, 102
Sexual dimorphism, 105–6, 120–21, 126, 132, 138, 157–58; in *A. afarensis*, 131–32, 133, 135
Simpson, George Gaylord, 40, 41, 44
Sinanthropus pekinensis (Peking man), 84–86
Sivapithecus, 129, 130
Size dimorphism, 131–32, 138
Skhūl (site), 153, 154, 159
Smith, Adam, 33
Smith, Grafton Elliot, 78, 79, 81
Social change, *see* Cultural change
Social Darwinism, 172
Social organization, 15, 23–24, 25–26, 158
Social philosophy, 163

Societies: change in, 30, 163, 164; denial of human status to others by, 19–20
Sociobiology, 16, 23–27, 161–62, 176, 178; is anachronism, 179
Speciation, 48–53, 127, 142, 155, 177; as adaptation, 43; following evolutionary crises, 59–61; key to evolutionary change, 62–65, 178–79; and morphological differentiation, 119–20; possibility of, in modern man, 186–87; and technological advance, 157
Species, 3, 37, 42–43, 159; as ancestors and descendants, 35–36, 48, 49; are discrete entities, 44–48, 49, 53, 178; extinction of, 58–59, 60; fixity of, 31–32, 34, 35–36, 44–45, 46, 47, 48 (*see also* Stasis; fragmentation of, 180–81, 186; new, 35–37, 157, 178–79; number of, 42–43, 120–28, 131, 133, 144, 179; recognition of, 119–20, 127–28, 135, 157 (*see also* Fossil record); stem, 117, 126; variability in, 131–32
Species selection, 63–64
Speech, 72
Spencer, Herbert, 33
Spy (Belgium), 70
Stadials, 77
Stasis (stability), 2, 3, 8, 61, 134, 156; cultural, 4, 164–65, 184–85; in Egyptian history, 177; in fossil record, 53, 58, 140–41; in *Homo erectus*, 144–45; interrupted by rapid change, 161–73; isolation and, 181; in species, 46, 47–48, 159, 177; in technology, 183–84
Steinheim (site), 157
Sterkfontein (site), 87–89, 90–91, 92, 93, 121, 124, 134, 135, 138, 139, 148, 149; stone tools, 140
Stone tools, 9–10, 67–68, 71, 91, 104, 145; Hadar, 113–14; as hallmark of *Homo*, 140; introduction of, 136, 156–57; Olduvai Gorge, 93, 94–95, 99–100; *see also* Acheulean hand-axe culture; Mousterian stoneworking tradition; Oldowan stoneworking tradition; Tool making and use
Stoneworking industries, 157; East Turkana, 109–10
Stringer, Chris, 148–49
Survival of the fittest, 33, 172
Swanscombe (site), 149
Swartkrans (site), 91, 92, 93, 121, 124, 139; robust hominids, 140–41; stone tools, 140
Sylvester-Bradley, P. C., 60
Systematics, 37, 39, 40, 49
Systems, 15–16

Tabun (site), 153–54
Taieb, Maurice, 110
Tasaday (people), 186
Taung, 82, 86–87, 92, 93, 124, 134
Technological change, 4, 9, 30, 172, 173, 183–84; and speciation, 157
Technology, 8–12, 145; *see also* Stone tools
Teggart, F. J., 162–65, 166, 168, 173, 179, 180, 181, 184, 187; *Theory of History*, 163, 170–71
Time, 39–40, 42, 132; and hominid chronology, 126, 127, 128; in hominid fossil record, 128–29; species in, 36–37, 44–45, 48
Tool making and use, 6, 8–12, 122, 129, 133, 134, 135, 152, 158; retained by *Homo sapiens*, 179; *see also* Stone tools
Trinil faunal zone, 143
Tuffs, 102, 104, 113; dating of, 98–99
2001 (movie), 6
Tylor, Edward, 22

Unas (king of Egypt), 165, 166
Upper Paleolithic toolmaking industries, 154
Upright posture, 8, 74, 122, 131, 133, 144; and brain development, 90–91, 115–16

Variability, 144–45, 185–86; of modern man, 147–48; morphological, 152, 157–58

Velikovsky, Immanuel, 23
Verteszöllös (site), 149
Vindija (site), 154
Virchow, Rudolf, 70, 74–75, 76

Walker, Alan, 142
Weidenreich, Franz, 85, 144
Weiner, Joseph, 80
White, Tim, 117, 126, 131, 132, 133
Wittfogel, Karl, 167

Woodward, Arthur Smith, 78–79
World views, 2–3
Wright, Sewall, 40–41
Writing, 161
Würm (glacial), 77

Zdansky, Otto, 83–84
Zinjanthropus boisei, 95–96, 97, 103, 104–5, 122, 123, 125, 141

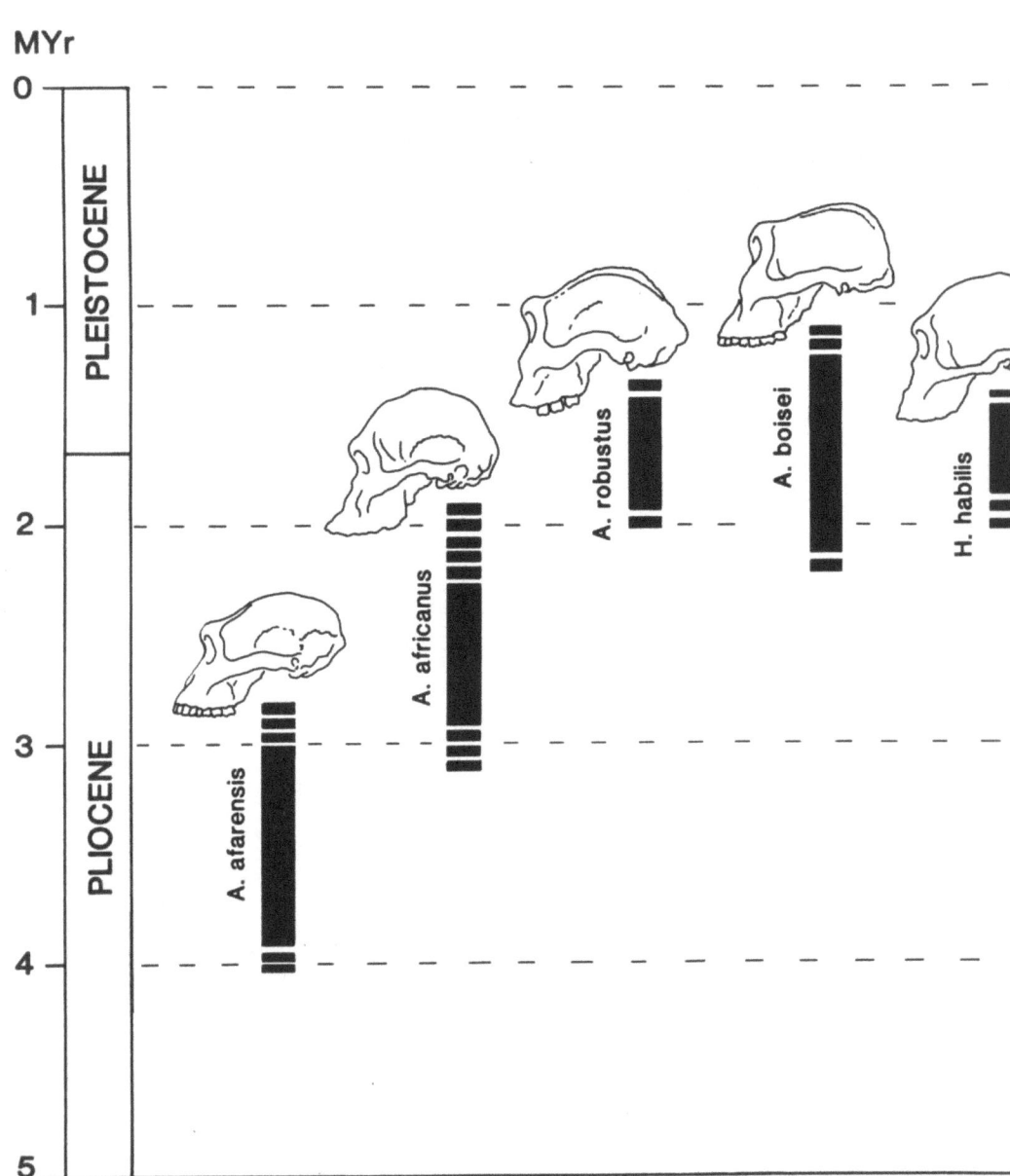

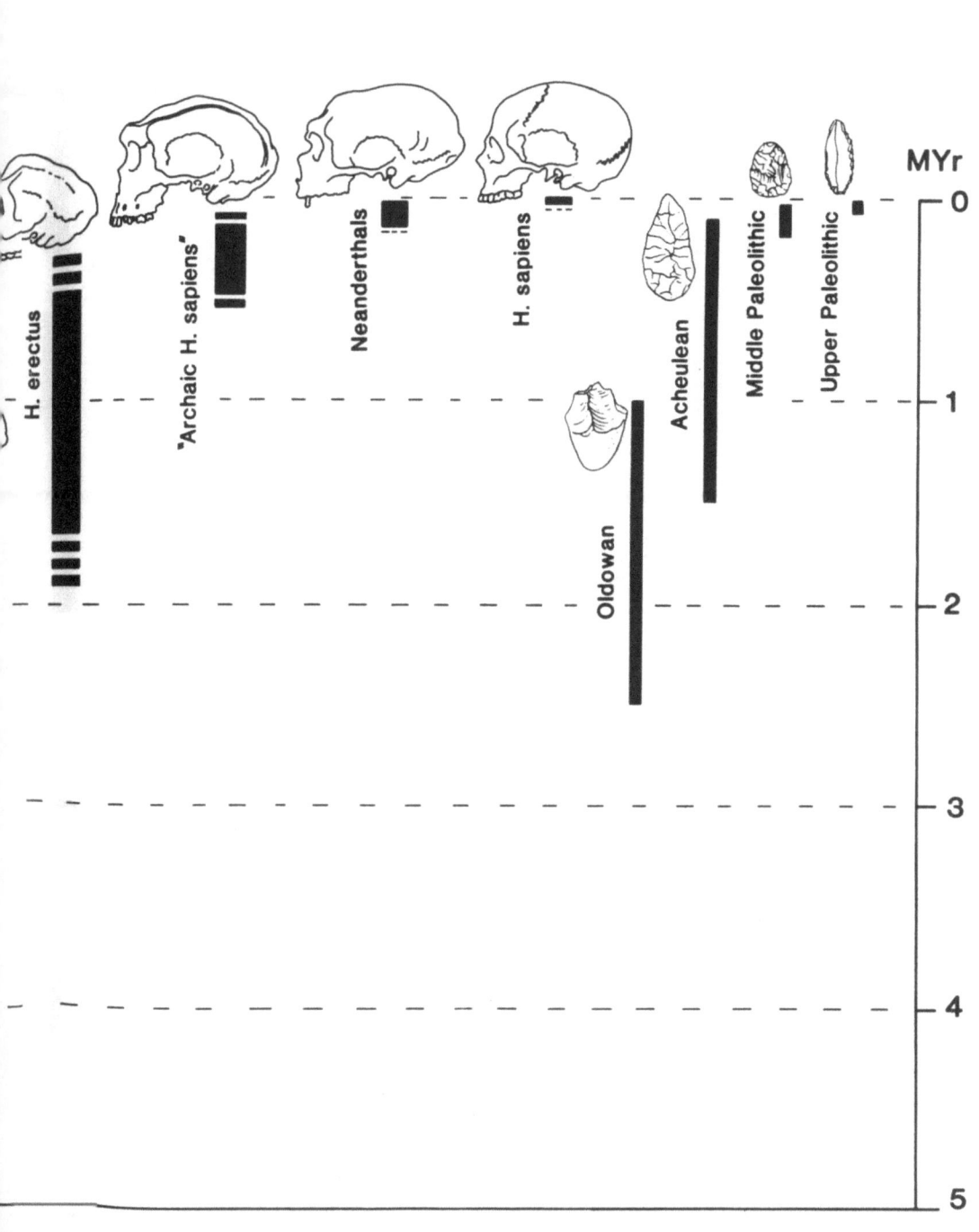

Bei Fragen zur Produktsicherheit wenden Sie sich bitte an:
If you have any questions regarding product safety,
please contact:

Walter de Gruyter GmbH
Genthiner Straße 13
10785 Berlin
productsafety@degruyterbrill.com